# CONDUIT
## BENDING AND FABRICATION

AMERICAN TECHNICAL PUBLISHERS, INC.
HOMEWOOD, ILLINOIS 60430-4600

IN PARTNERSHIP WITH NJATC

*Conduit Bending and Fabrication* contains procedures commonly practiced in industry and the trade. Specific procedures vary with each task and must be performed by a qualified person. For maximum safety, always refer to specific manufacturer recommendations, insurance regulations, specific job site and plant procedures, applicable federal, state, and local regulations, and any authority having jurisdiction. The material contained is intended to be an educational resource for the user. Neither American Technical Publishers, nor the National Joint Apprenticeship & Training Committee for the Electrical Industry is liable for any claims, losses, or damages, including property damage or personal injury, incurred by reliance on this information.

American Technical Publishers, Inc., Editorial Staff

Editor in Chief:
> Jonathan F. Gosse

Production Manager:
> Peter A. Zurlis

Art Manager:
> James M. Clarke

Technical Editor:
> Eric F. Borreson

Copy Editor:
> Catherine A. Mini

Cover Design:
> Mark S. Maxwell

Illustration/Layout:
> Mark S. Maxwell
> Peter J. Jurek
> Jennifer M. Hines
> Thomas E. Zabinski

CD-ROM Development:
> Carl R. Hansen
> Christopher J. Bell
> Peter J. Jurek

1 2 3 4 5 6 7 8 9 – 07 – 15 14 13 12

Printed in the United States of America

ISBN 978-0-8269-1267-1

 This book is printed on recycled paper.

 16

# Acknowledgments

Technical information and assistance was provided by the following companies, organizations, and individuals.

Carlon
Detroit Electrical JATC
IBEW-NECA Technical Institute (Alsip, IL)
Klein Tools, Inc.
Library of Congress
National Park Service
Ridge Tool Company
Steel Tube Institute of North America
USA Speedshore Corporation

NJATC Acknowledgments
Technical Editor
       William R. Ball, NJATC Staff

Technical Reviewers
       Tom Bowes, Detroit Electrical JATC
       Tony Griffin, IBEW-NECA Technical Institute (Alsip, IL)
       Harold Ohde, IBEW-NECA Technical Institute (Alsip, IL)
       Jim Paladino, Omaha Joint Electrical Apprenticeship and Training Committee

# Table of Contents

**1**

**Raceways and Conduit Systems** ...................... **1**

Raceways and Conduit ..................................................................2

Conduit and the NEC® .................................................................4

Summary .....................................................................................9

**2**

**Hand Bending—90° Bends** ........................... **11**

Tools .........................................................................................12

Basic Bends ..............................................................................17

Bend Pre-Positioning .................................................................23

Application—Three-Bend Back-to-Back Bends ...........................25

Summary ...................................................................................27

**3**

**Hand Bending—Offset Bends and Kicks** ........ **29**

Offset Bends ..............................................................................30

Kicks .........................................................................................52

Application—Parallel Offsets with Conduit of Different Sizes ...........54

Summary ...................................................................................56

## Hand Bending—Saddles and Corner Offsets .. 57

Saddles ...............................................................................58

Three-Bend Saddles ..............................................................61

Four-Bend Saddles ...............................................................66

Corner Offsets .....................................................................69

Compound 90° Bends ............................................................70

Application—Three-Bend Saddle around a Drain .......................73

Summary ............................................................................76

## Mechanical and Electric Benders ...................77

Mechanical Benders .............................................................78

Layout ...............................................................................81

Fabrications ........................................................................88

Electric Benders ..................................................................99

Summary ..........................................................................104

## Hydraulic Benders .......................................105

Hydraulic Principles ...........................................................106

Hydraulic Benders ..............................................................106

Layout and Fabrication .......................................................109

Summary ..........................................................................114

# Table of Contents

**7** **Other Conduit Types**.................................. **115**

Metallic Conduit ...................................................116

PVC-Coated Conduit ............................................117

PVC Conduit .......................................................120

Summary ............................................................123

**8** **Threaded Conduit**................................. **125**

Threaders ...........................................................126

Threaded Conduit.................................................133

Summary ............................................................137

**9** **Advanced Bending Technique** ..................... **139**

Segmented Bends ................................................140

Concentric Bends.................................................147

Application—Arched Ceiling ...................................152

Summary ............................................................154

## 10 Underground Conduit Installation Procedures .. 155

Job Planning...........................................................156

Surveying Instruments ...........................................158

Excavation.............................................................164

Underground Installation Procedures......................169

Backfilling .............................................................171

Summary...............................................................174

## A Appendix .................................................... 175

Appendix A ...........................................................177

Appendix B ...........................................................183

Appendix C ...........................................................197

## Glossary ............................................... 207
## Index .................................................... 209

## CD-ROM Contents

- *Using this CD-ROM*
- *Quick Quizzes™*
- *Illustrated Glossary*
- *Bending Calculator*
- *Procedural Videos*
- *Reference Material*

# Introduction

## CONDUIT BENDING and FABRICATION

*Conduit Bending and Fabrication* provides a comprehensive overview of conduit bending and fabrication procedures and methods. This textbook is designed to develop basic competencies in electrical apprentices and beginning learners.

*Conduit Bending and Fabrication* begins with a thorough discussion of hand bending tools, conduit layout, and simple hand bending procedures. Subsequent chapters expand on the basics to include procedures for hand bending offsets and saddles. After the principles of bending and layout are covered, later chapters cover the same types of bends made with mechanical, electric, and hydraulic benders.

*Conduit Bending and Fabrication* contains 10 chapters. At the beginning of each chapter, a chapter Table of Contents makes it easy to find desired information. At the end of each chapter, a chapter Summary helps learners review key concepts and reinforce common procedures.

Key terms are italicized and defined in the text for additional clarity. Tech Facts and vignettes throughout the text provide information that enhances text content. An extensive Glossary and Appendix provide useful, easy-to-find information. A comprehensive Table of Contents and Index simplify navigation and make finding information easy.

The *Conduit Bending and Fabrication* CD-ROM located at the back of the book is designed as a self-study aid to complement information presented in the book and includes Quick Quizzes™, an Illustrated Glossary, Video Clips, and Reference Material. The Quick Quizzes™ offer an interactive review of topics in a chapter. The Illustrated Glossary provides a helpful reference to terms commonly used in industry. The Video Clips are a collection of video clips and animated graphics showing common conduit bending procedures. Reference Material provides access to Internet links to manufacturer, association, and American Tech resources. Clicking on the American Tech web site button (www.go2atp.com) or the American Tech logo accesses information on related electrical training products.

*The Publisher*

# Features

## CONDUIT BENDING and FABRICATION

Power-driven threaders are high-torque tools and can cause serious injury if not used correctly. Many manufacturers provide a safety bracket that mounts onto conduit to prevent the threader from moving. In addition, some threaders have a boss or hub in the end of the handle where a short piece of 1" rigid conduit can be inserted or screwed for additional leverage to help control the tool.

Power-driven threaders are often used on jobs where a lot of threading is done but where there may not be space available to set up a fabrication shop. They are also portable enough to be used on projects that require the electrician to move many times throughout the workday.

### Large Power Threaders

Large power threaders are different from hand- and power-driven threaders. Instead of the rotating die head found on handheld threaders, power-driven threaders have a fixed head and rotate the conduit. Large power threaders require a stand or bench to hold the tool. They can be very dangerous due to the amount of torque they produce. The most common type of large power threader is the threading machine.

**Threading Machines.** Threading machines are available to thread conduit from ½" to 4". Threading machines range from simple threaders that have manual oilers but do not have cut-length-limiting devices, to complex threaders with automatic oilers and devices to cut standard thread. The simplest setup uses a threading machine with a hand threader. Die heads on threading machines often are adjustable for varying sizes of conduit. Typically, one die head can be adjusted to cut threads on a particular range of conduit sizes. **See Figure 8-9.**

A threading machine has chucks located at the rear and front of the threader. When conduit is placed in the threader, the chuck at the rear is closed first. The main chuck at the front is closed last. A threading machine setup usually includes a conduit cutter that is attached to the threading machine stand. Conduit that is fastened to the main chuck of the threading machine can be cut to size by using the conduit cutter. **See Figure 8-10.**

**Tech Fact**

Rigid metal conduit and intermediate metal conduit are typically threaded in order to attach the conduit to fittings, attach lengths of conduit together, and to mechanically attach the conduit to junction boxes, troughs, and panels. Threaded conduit connections are often a vital part of the grounding system and must be fabricated properly in order to do their job. In hazardous locations, the threads play a very important part in the prevention of explosions that might otherwise be caused by certain types of electrical equipment. It is therefore imperative that an electrician know how to properly thread these types of conduit.

**Threading Machines**

ADJUSTABLE DIE HEAD

Cutter
Chucks
Die head
Reamer
Carriage
Ridge Tool Company

THREADING MACHINE

**Figure 8-9.** A threading machine is often used for threading conduit. Many include adjustable die heads to simplify their setup.

*Tech Facts provide background information of interest to electricians.*

*Close-up photographs illustrate details of equipment operation.*

---

**Finding the Center of a Bend**

If the bender does not have a benchmark for the center of the bend, or if a 30° angle is chosen, a new benchmark can be placed on the bender. The procedure to find the center of a bend and create a new benchmark is as follows:

1. Make a pencil mark on the conduit at any convenient location.

2. Align any convenient benchmark with the pencil mark and fabricate a 30° bend on a scrap piece of conduit.

3. Find the center of the bend by extending (with pencil marks) both of the inside edges of the conduit through the bend. The point at which the two marks intersect is the center of the bend.

4. Place the bender back on the conduit and align the original benchmark with the same pencil mark as before. Place a permanent marking at the center of the bend for the 30° bends on the bender at the point where the center of the bend mark touches the shoe.

1. Place pencil mark.   2. Align pencil mark with arrow.

3. Find center of bend.   4. Transfer mark to bender.

*Step-by-step procedures describe proper methods of bending conduit.*

*Equipment from leading manufacturers is depicted throughout the text.*

---

# Hand Bending— Offsets and Kicks

### CONDUIT BENDING and FABRICATION

**3**

OFFSET BENDS ........................................................................ 30
KICKS ...................................................................................... 52
APPLICATION—PARALLEL OFFSETS WITH CONDUIT OF DIFFERENT SIZES ...... 54
SUMMARY ............................................................................... 56

After 90° bends, offset bends are the most common bend. Offset bends may be used to change elevation, to get around an obstacle such as a structural member, another conduit, or ductwork, or to enter a junction box or an enclosure.

An understanding of basic trigonometry is useful in conduit bending. There are several angles that are commonly used in bending offsets. The most common angles for offset bends are 22½°, 30°, and 45°.

When an offset needs to be bent in a piece of conduit, a right triangle can be used to represent the bend. The triangle relates the angle of the bend and the offset. The triangle allows the use of trigonometry to find the distance between the two bends.

There are many situations where the bends must fall in a specific location on the conduit. This is often the case when conduit must be worked around an obstruction or when the bends of several conduits must be aligned to obtain a neat and workmanlike result. The total run length of conduit is reduced when an offset bend is made. This shrink must be taken into account when locating the bend points. Parallel offsets present unique challenges. Special techniques are used to keep the spacing even through the angled sections of the bends.

#### OBJECTIVES

1. Define offset bend and kick and distinguish between them.
2. Demonstrate the multiplier method for making offset bends.
3. Calculate the distance between bends.
4. Describe shrink and explain how to calculate shrink.
5. Calculate the shift required when bending parallel offsets.
6. Demonstrate the measured rise method for making offset bends.
7. Demonstrate how to use shrink to calculate the required length of conduit before bending.

**29**

*A chapter Table of Contents makes it easy to find relevant information.*

*A chapter Introduction provides an overview of key content found in the chapter.*

*Step-by-step photographs coordinate with procedures and show proper methods.*

*Objectives provide a summary of learning goals for the chapter.*

# About the Author

Steve Lipster has more than 20 years of experience as an electrician and JATC instructor. He completed his apprenticeship at the Columbus Joint Apprenticeship and Training Committee. One year after completing his apprenticeship, he was appointed as a labor trustee on the same committee that sponsored his apprenticeship. Within a short time, he began teaching at the JATC. He authored the JATC publications *Power Harmonics, Surveying and Excavation for Wireman,* and *Uninterruptible Power Supplies.*

He was awarded the prestigious Founders Scholarship from the International Brotherhood of Electrical Workers and returned to college to pursue a degree in Technology Education at The Ohio State University, where he graduated magna cum laude. Shortly before his graduation, he was appointed JATC/Electrical Trades Center director. In addition to his membership in the IBEW, Mr. Lipster is also a member of the National Fire Protection Association and the Institute of Electrical and Electronics Engineers, and is currently the Chairman of the Board of Directors of the Heart of Ohio College Tech Prep Consortium.

# Raceways and Conduit Systems

## CONDUIT BENDING and FABRICATION

**1**

RACEWAYS AND CONDUIT ........................................................................................ 2
CONDUIT AND THE NEC® ........................................................................................ 4
SUMMARY ................................................................................................................... 9

*C*onduit as a practical raceway for electrical systems has been in use from the beginning of the electrical industry. Although the decades since the introduction of conduit have produced many alternative wiring methods, conduit offers a certain amount of flexibility and unmatched physical protection for conductors.

Early conduit designs included zinc tubes and spiral-wound paper tubes. These conduit designs were difficult to install because they could not be bent in the field. The development of lined iron gas pipe led to improvements in installation procedures because, to some extent, the iron pipe could be bent in the field.

The development of rubber insulation for conductors made the lined iron gas pipe conduit systems obsolete and unlined iron gas pipe was used. The development of enamel-coated steel conduit led to further improvements.

Rigid metal conduit is a threadable conduit with fairly thick walls. Intermediate metal conduit (IMC) is a threadable conduit with walls of intermediate thickness. Electrical metallic tubing (EMT) is a nonthreadable conduit with thinner walls than rigid or IMC. Because of its lighter weight and lack of threads, there are several restrictions on the use of EMT.

## OBJECTIVES

1. Briefly outline the history of raceways and conduit.
2. Describe the differences between rigid metal conduit and electrical metallic tubing.
3. Summarize the appropriate section of the NEC® for rigid, EMT, PVC, and IMC.
4. Describe the color coding of the thread protector caps used with rigid conduit and IMC.

*Library of Congress*

## RACEWAYS AND CONDUIT

The National Electrical Code® (NEC®) recognizes many different types of raceways for the protection and routing of electrical conductors. While some take the form of channels, troughs, gutters, or ducts, the most common raceways consist of conduit or tubing. Conduits such as rigid metal conduit (RMC), intermediate metal conduit (IMC), and rigid nonmetallic conduit (RNC) can be used in conjunction with electrical metallic tubing (EMT) and other raceways to form a system that allows conductors to be drawn into them after the conduit runs are completed.

When the NEC® uses the word conduit, it means only those raceways that contain the word "conduit" in their titles. However, by common usage, electrical metallic tubing (EMT) and other raceways are also called conduit.

Conduits are typically bundled for shipment in groups of 5 or 10 pieces. For large shipments, these bundles are banded onto large skids that may have as many as 50 bundles. Caution should be used when breaking these large shipping bundles open. The individual bundles can shift and fall, causing severe personal injury.

### Early Wiring Methods

In 1879, Thomas Edison demonstrated the first practical incandescent lamp. **See Figure 1-1.** In many ways, developing the electric lamp was a major step in creating the electrical industry. Edison knew he would not be able to exploit this achievement unless methods of generating, distributing, and using electricity were developed.

Historians generally say that Edison's Manhattan Pearl Street generating station was the first practical commercial electrical system. **See Figure 1-2.** The Pearl Street generating station was built by the Edison Electric Illuminating Company in 1882 and was powered by a steam-powered DC generator. It originally provided power to about 50 buildings.

Eventually, the early Edison DC power systems were replaced by AC systems invented by Nikola Tesla and promoted by Westinghouse. The AC systems allowed for efficient transmission of power over greater distances. The expansion of the Westinghouse AC systems led to the increased use of electricity.

Architects designing buildings wanted the ability to install electrical power systems during construction. However, wire insulation was very poor, generally consisting of a braided fabric with a wax binder. This insulation deteriorated quickly and caused fires. In addition, the exposed wiring had the potential to cause personal injury from electrical shock.

From an aesthetic and safety point of view, it made sense to install the wires behind walls. Building owners needed the ability to protect people from the wires. However, the owners still needed to be able to replace the wires frequently to prevent fires. These problems led to the development of conduit to protect the wires and to make it easier to replace damaged wires.

*T*ech Fact

Conduit fittings, such as connectors and elbows, are the weak points of a raceway system. Fittings may loosen or corrode and destroy the grounding path.

### Edison's Incandescent Lamp

*National Park Service*

**Figure 1-1.** The electric lamp was a major step in creating the electrical industry.

*Thomas Edison was very important in the development of the commercial electric industry. His development of the incandescent light bulb led to the increased demand for electricity.*

## Early Conduit Designs

Conduit was present from the very beginning of the electrical industry. The electrical mains for customers of the Pearl Street generating station were installed underground. The mains were constructed of a rod conductor installed in an iron pipe. The pipe was filled with an asphalt-like substance to waterproof the conduit and act as an insulator.

Taps were made from these mains leading to the customers. Once inside the building, wires were installed on wooden cleats and connected to newly designed switches, lampholders, and other devices.

**Zinc Tubes.** In the 1880s, electricians began experimenting with lightweight zinc tubes to protect wires installed behind walls. These zinc tubes were very light and could not be bent on the job. However, the zinc tubes did provide a method of concealing the wires and allowed new wires to be pulled to replace deteriorated wires without damaging building finishes. One zinc tube was used for each wire.

***Figure 1-2.*** *The Pearl Street generating station was the first practical commercial electrical system.*

**Spiral-Wound Paper Tubes.** Shortly after the development of the zinc tubes, a conduit system was introduced that used spiral-wound paper tubes covered with asphalt. This system included brass couplings and long-radius 45° and 90° elbows. The advantage of this system was that the conductor was protected within a raceway that was a better insulator than the wire insulation. A major disadvantage of this system was that it was difficult to install. In addition, the conduit had to be protected from moisture to prevent any deformation or other damage in the paper conduit.

**Iron Gas Pipe.** In the 1890s, iron gas pipe lined with wood, fiber, or paper became available. As with the paper conduit systems, long-radius 45° and 90° elbows were available. The couplings and fittings were threaded to allow a strong, durable connection. This iron gas pipe conduit system provided the best protection yet for the enclosed wires as well as a reliable ground path. It could be field bent to optimize routing, as long as the bends were kept to a small angle with a large radius.

The practice of using one conduit for each conductor was discarded when iron conduit was developed. Large, induced magnetic fields are produced when a single alternating current conductor is installed in an iron raceway. This reduces the current carrying capacity of the conductor. Not only did installing multiple conductors in a single conduit make economic sense, it was required to improve the efficiency of an electrical distribution system.

In the 1890s, the first viable rubber wire insulation was developed, making the lined iron gas pipe conduit unnecessary. A rubber compound was vulcanized to the wire to act as a dielectric. Linen fabric was braided around the rubber to protect the insulation from abrasion while the wire was being pulled into the conduit. While this type of insulation was not as good as modern insulation, it lasted for the life of the system.

**Steel Conduit.** Steel conduit was introduced at about the same time as insulated wire. Steel is more ductile than iron and can be bent with a fairly short bend radius and large bend angles. The steel conduit of the late 19th century was very different than modern conduit. One difference was that the early steel conduit had a heavier wall thickness than modern rigid conduit because it was manufactured to gas pipe standards.

Another major difference was the appearance of the conduit. While galvanizing was available at that time, it was not very effective. The older galvanizing methods left sharp edges on the conduit that would have damaged the wire as it was being pulled into the conduit. To avoid this, steel conduit of the day was enameled both inside and out. In about 1903, an effective electro-galvanizing process was developed and galvanized steel conduit quickly became the industry standard.

---

*Tech Fact*

There are many sections of the NEC® that describe the uses of conduit and raceways. One of the oldest sections says that there shall not be more than the equivalent of four quarter bends (360° total) between pull points, for example, conduit bodies and boxes. This is true for all types of conduit and tubing used as raceways.

---

## CONDUIT AND THE NEC®

There are several sections of the NEC® that refer to conduit and raceways. Each of these raceways has its own article in the NEC®. The sections entitled "Uses Permitted" and "Uses Not Permitted" should be reviewed for the particular raceway being considered. Some of these articles require the raceway to be listed. If that is the case, the raceway shall be installed in accordance with any instructions included in the listing or labeling.

### Rigid Metal Conduit (RMC)

*Rigid metal conduit (RMC)* is a threadable conduit with fairly thick walls. RMC is commonly called "rigid" in the field. In the early part of the 20th century, rigid steel conduit and galvanized rigid steel conduit were the only conduit systems allowed by the NEC®. Rigid is discussed in NEC® Article 344. Rigid is permitted under all atmospheric conditions and in all types of occupancies. When installing rigid, dissimilar metals that could cause galvanic action should be avoided.

Rigid is generally available in 10′ lengths with a coupling on one end. The NEC® allows lengths shorter or longer than 10′ to be shipped. However, it is relatively uncommon for a manufacturer to provide lengths other than the standard 10′ size. Therefore, prices for these lengths can be expected to be higher than for standard lengths.

Rigid is available in sizes from ½″ in diameter up to 6″ in diameter. **See Figure 1-3.** Sizes may be given in inches or millimeters. In addition, ⅜″ rigid may be used between a motor and its junction box when the junction box is not part of the motor housing.

**Thread Protector Caps.** Rigid is shipped with colored protector caps to protect the threads. The colored protector caps are blue for sizes 1″, 2″, 3″, 4″, 5″, and 6″. The caps are black for sizes ½″, 1½″, 2½″, and 3½″. The caps are red for sizes ¾″ and 1¼″. **See Figure 1-4.** In some parts of the country, the color code may also be found on the nylon straps used to bundle the conduit.

The color of the protector caps is an added feature that aids in recognizing the size of the conduit. If at all possible, the caps should be left on the threads until the bending is complete. This provides added protection for the threads.

### Electrical Metallic Tubing (EMT)

*Electrical metallic tubing (EMT)* is a lightweight tubular steel raceway without threads on the ends. In the 1928 revision to the NEC®, EMT was first approved in a very limited scope. EMT is discussed in NEC® Article 358. It is the lightest metallic raceway available. The lengths are typically joined together with setscrew couplings or compression fittings. **See Figure 1-5.**

---

**Tech Fact**

According to the NEC®, a bushing shall be provided where a conduit enters a box, fitting, or other enclosure to protect the wire from abrasion, unless the design of the box, fitting, or enclosure is such as to afford equivalent protection.

---

### Rigid Thread Protector Caps

| Color | Sizes | Examples |
|-------|-------|----------|
| Blue | Inch sizes | 1″, 2″, 3″, 4″, 5″, 6″ |
| Black | ½″ sizes | ½″, 1½″, 2½″, 3½″ |
| Red | ¼″ sizes | ¾″, 1¼″ |

**Figure 1-4.** Thread protector caps protect the threads and use color coding to help with identification.

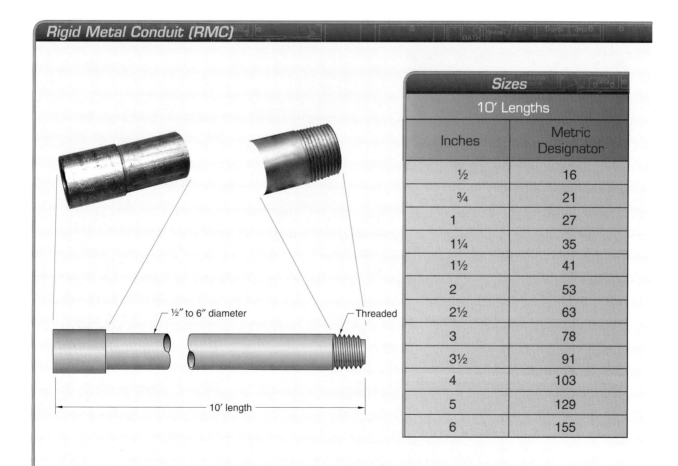

### Rigid Metal Conduit (RMC)

½″ to 6″ diameter

Threaded

10′ length

| Sizes | |
|-------|-------|
| **10′ Lengths** | |
| Inches | Metric Designator |
| ½ | 16 |
| ¾ | 21 |
| 1 | 27 |
| 1¼ | 35 |
| 1½ | 41 |
| 2 | 53 |
| 2½ | 63 |
| 3 | 78 |
| 3½ | 91 |
| 4 | 103 |
| 5 | 129 |
| 6 | 155 |

**Figure 1-3.** *Rigid metal conduit (RMC) is available in sizes from ½″ to 6″ in diameter with threaded ends.*

## Electrical Metallic Tubing (EMT)

**COMPRESSION COUPLING**

**SETSCREW COUPLING**

**COMPRESSION CONNECTOR**

**SETSCREW CONNECTOR**

### Sizes

#### 10' Lengths

| Inches | Metric Designator |
|--------|-------------------|
| ½ | 16 |
| ¾ | 21 |
| 1 | 27 |
| 1¼ | 35 |
| 1½ | 41 |
| 2 | 53 |
| 2½ | 63 |
| 3 | 78 |
| 3½ | 91 |
| 4 | 103 |

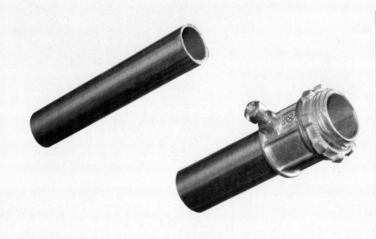

**EMT SHALL NOT BE USED**

- Where subject to severe physical damage
- In corrosive atmospheres
- Where buried in cinder fill
- In hazardous locations (with exceptions)
- For support of equipment except conduit bodies no larger than largest trade size of EMT
- Where practicable, in contact with dissimilar metals

**EMT IS AVAILABLE**

- In 10' lengths
- In ½" to 4" diameters

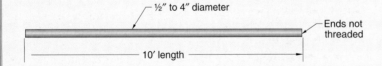

½" to 4" diameter

Ends not threaded

10' length

**Figure 1-5.** *Electrical metallic tubing (EMT) is unthreaded thinwall tubing. The lengths are typically joined together with setscrew couplings or compression fittings.*

EMT has a much thinner wall thickness than rigid and does not require threading. Therefore, it supplies a degree of protection to the conductors while being cost effective. EMT is about 40% thinner in wall thickness than rigid and is commonly called "thinwall" in the field. There are six restrictions on the use of EMT. It cannot be used in the following situations:

- where subject to severe physical damage
- in corrosive atmospheres
- where buried in cinder fill
- in hazardous locations, except as specifically permitted
- for the support of fixtures or other equipment except conduit bodies no larger than the largest trade size of the EMT
- where practicable, in contact with dissimilar metals

### Rigid Nonmetallic Conduit (RNC)

*Rigid nonmetallic conduit (RNC)* is a conduit made of materials other than metal. RNC is discussed in NEC® Article 352. These materials are recognized as having suitable physical characteristics for direct burial in the earth. One of the more common types of RNC is polyvinyl chloride (PVC). **See Figure 1-6.**

Originally, PVC water pipes were given a number designation depending on the wall thickness. The thinnest wall thickness was designated as 10. The next thicker size was 20, then 30, 40, and up to 100. When PVC was presented to the electrical industry, the three sizes adopted were Schedule 20, 40, and 80.

PVC is waterproof, rustproof, and does not corrode in most locations. Because Schedule 40 PVC does not have the strength of metal conduits, it cannot be used where subject to physical damage. Schedule 80 PVC has a heavier wall thickness and is suitable for most locations where RMC is used. PVC must be supported as required in the NEC®. If the length change in a run of PVC exceeds ¼″ due to thermal expansion, an expansion fitting must be provided.

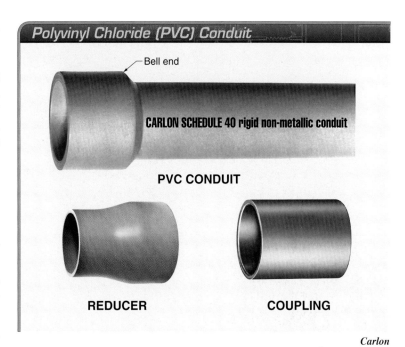

*Carlon*

**Figure 1-6.** *PVC conduit is a common type of rigid nonmetallic conduit. PVC is available in thinwall and thickwall grades with many types of couplings and fittings.*

**Permitted Uses.** RNC, including PVC, was first approved in the 1960s. There are many permitted uses listed in the NEC®. Because of its nonmetallic, noncorrosive nature and ease of installation, PVC conduit is commonly used in underground installations. It is also used aboveground for carrying and protecting grounding electrode conductors. Because of the nonmetallic nature of PVC, grounding conductors can carry extremely high fault currents safely without inducing the large magnetic fields that would be found in a steel conduit raceway system.

As a general note, the use of PVC in extremely cold environments should be carefully evaluated. Extreme cold can make PVC brittle and therefore susceptible to physical damage. PVC is one of the few raceways that can be installed in cinder fill without the use of additional protective measures. It is also permitted for exposed work if it is identified as sunlight resistant and will not be subject to physical damage.

**Uses Not Permitted.** The NEC® lists the prohibited uses for rigid nonmetallic conduit.

Temperature considerations should always be considered when selecting RNC for a specific application. In general, RNC should not be installed where there are significant temperature extremes. RNC is prohibited from use where it is subjected to ambient temperatures in excess of 122°F and RNC is only permitted in hazardous locations under very specific conditions.

### Intermediate Metal Conduit (IMC)

*Intermediate metal conduit (IMC) is a raceway of circular cross-section with an intermediate wall thickness designed for protection and routing of conductors.* IMC is discussed in NEC® Article 342. The NEC® requires that IMC and all associated fittings, elbows, and couplings be listed. The NEC® has specific requirements when IMC is installed in corrosive environments or is in indirect contact with cinder fill.

The wall thickness of IMC is between the wall thicknesses of EMT and rigid conduit. IMC is available in trade sizes ½″ through 4″. Like rigid conduit, IMC is generally available in 10′ lengths with a coupling on one end. It is made of steel, is threadable, and is frequently used because of its excellent protective qualities, large internal diameter, and ease of installation.

IMC is composed of a less-ductile grade of steel than rigid. This can cause some problems when bending IMC. It is not unusual for IMC to split when bent against the seam. This conduit also has much more springback than other metal conduits and requires adjustment when bending to specific angles.

**Thread Protector Caps.** IMC is shipped with colored protector caps to protect the threads. The colored protector caps are orange for sizes 1″, 2″, 3″, and 4″; yellow for sizes ½″, 1½″, 2½″, and 3½″; and green for sizes ¾″ and 1¼″. **See Figure 1-7.**

---

### Tech Fact

A bushing is a fitting placed on the end of a conduit to protect the conductor's insulation from abrasion.

---

| IMC Thread Protector Caps | | |
|---|---|---|
| Color | Sizes | Examples |
| Orange | Inch sizes | 1″, 2″, 3″, 4″ |
| Yellow | ½″ sizes | ½″, 1½″, 2½″, 3½″ |
| Green | ¼″ sizes | ¾″, 1¼″ |

**Figure 1-7.** *Thread protector caps protect the threads and use color coding to help with identification.*

## SUMMARY

- Conduit has been used from the beginning of the electrical industry.

- Early conduit designs included zinc tubes, spiral-wound paper tubes, and iron gas pipe.

- The development of rubber insulation for conductors and galvanized steel conduit made the lined iron gas pipe systems obsolete.

- Galvanized steel conduit can be bent in the field to improve installation procedures.

- Rigid conduit is a threadable conduit with fairly thick walls.

- Rigid can generally be used in all atmospheric conditions and all types of occupancies.

- Color-coded thread protector caps help protect the threads and help identify the size.

- EMT has thinner walls than rigid and does not require threading.

- EMT has several restrictions on its use.

- PVC conduit is waterproof, rustproof, and does not corrode in most locations.

- IMC is a conduit with intermediate wall thickness.

# Hand Bending — 90° Bends

## CONDUIT BENDING and FABRICATION

**2**

TOOLS ............................................................................................12
BASIC BENDS .................................................................................17
BEND PRE-POSITIONING ..............................................................23
APPLICATION – THREE-BEND BACK-TO-BACK BEND...........................25
SUMMARY .......................................................................................27

A skilled electrician must be able to efficiently make conduit bends using a hand bender and hand tools. Hand benders are available to bend a variety of sizes of conduit, ranging from ½″ up to 1¼″ EMT or 1″ rigid.

All benders have similar components and markings that must be understood in order to make accurate bends. The various markings are used to align the correct part of a bender with a pencil mark on the conduit. The ability to read a tape measure or rule is very important because conditions in the field determine the type and size of bends to be made for a particular job.

The most common hand bends are stub-up, 90° bends, and back-to-back 90° bends. Fabricating these bends is an essential skill that must be mastered in order to become proficient at bending all types of conduit.

A stub-up bend requires the use of the bender take-up so that the stub extends the correct length from the back of the bend. A back-to-back bend requires the use of the bender gain when conduit must be cut and threaded before bending.

### OBJECTIVES

1. List the components of a hand bender.
2. List the shoe markings on hand benders.
3. Explain the use and practical limitations of hickeys.
4. Explain how to determine the take-up of a bender.
5. Demonstrate the basic arithmetic and take-up used when fabricating 90° bends.
6. Calculate the amount of gain in 90° bends and back-to-back 90° bends.

## TOOLS

There are several tools required when making hand bends in conduit. Standard hand benders are used to make bends in conduit up to 1¼″ EMT. Hickeys are used to make bends in rigid conduit. Conduit reamers are used to remove any burrs from cut edges. Tape measures and rules are used to measure distances and lengths. Levels and protractors are used to check bends. Calculators are used in many applications.

### Hand Benders

Standard hand benders are available from several manufacturers and come in sizes that can bend ½″ to 1¼″ EMT conduit. **See Figure 2-1.** The NEC® specifies a minimum bend radius for field bends depending on the conduit diameter. Benders are manufactured to deliver that particular radius. Generally, a particular bender size is used to bend the corresponding size of conduit. However, there are times when a smaller conduit should be bent on a bender one size larger, such as when a larger bend radius is used to match bends between adjacent pieces of conduit. A larger bender may also be used to increase the radius of bends in order to ease wire pulling.

Rigid conduit may also be bent with EMT benders. However, the next larger size bender must be used. For example, ½″ rigid conduit can be bent with a ¾″ EMT bender that is also designed to bend rigid conduit. When bending rigid conduit with an EMT bender, care should be taken to prevent damage.

Occasionally a bender can be "sprung" when bending rigid conduit. This means that the hook becomes bent in relation to the shoe. If this occurs, the bender is ruined and should be thrown away.

**Bender Classification.** The two broad classifications of hand benders are plumb 30 benders and plumb 45 benders. These classifications are based on the angle generated when the bender handle is in the plumb (vertical) position. A plumb 30 bender produces a 30° angle when the handle is plumb. A plumb 45 bender produces a 45° angle when the handle is plumb. **See Figure 2-2.**

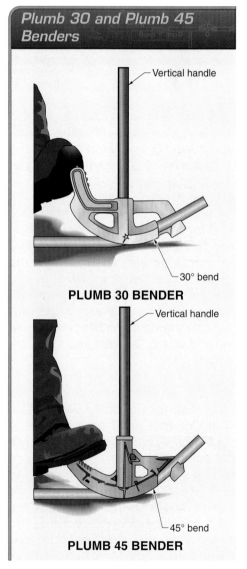

**Plumb 30 and Plumb 45 Benders**

Vertical handle
30° bend
**PLUMB 30 BENDER**
Vertical handle
45° bend
**PLUMB 45 BENDER**

*Figure 2-2. Different bender designs give different bends.*

**Hand Benders**

*Figure 2-1. Standard hand benders are available from several manufacturers and come in a variety of sizes and configurations.*

It should be noted that these statements are only approximately accurate. The handle of the bender must be moved past plumb before it is released. This allows for springback of the conduit. *Springback* is the property of conduit that causes it to unbend slightly after a bend is completed.

In addition, all benders are slightly different and the angle markings may be slightly off. Therefore, it is not sufficient to use a level to place the handle plumb to the floor. The conduit itself must be measured with a torpedo level or protractor level. An electrician must become familiar with the bender so that quick and accurate bends can be made.

Choosing a bender is often a matter of personal preference. The choice may also depend on the type of bender available at the job site. Since benders are sometimes used with a torpedo level, an electrician should consider the type of torpedo level that is best for the job. **See Figure 2-3.** In any case, it is a good idea to use the same bender whenever possible. Using the same bender allows the electrician to fabricate more consistent and accurate bends than if using different benders.

**Bender Components.** All benders have similar components. **See Figure 2-4.** Every bender has a shoe. A *bender shoe* is the curved part of a bender that forms the conduit during fabrication. The shoe is used to hold the conduit as it is wedged, or coined, to produce the bend.

A *bender handle* is a tube or lever used to hold the bender while in use. The handle can be factory made or it may be produced on the job by threading rigid conduit of the proper diameter and length into the bender shoe. A rule of thumb for handle length is that the combined length of the handle and bender should reach up to the electrician's elbow. It may be tempting to increase the length of the handle to gain mechanical advantage, but the extra handle length makes it cumbersome for many types of bends.

A *bender foot pedal* is the part of the bender where foot pressure is applied in order to bend the conduit. The foot pedal is designed to maximize the amount of foot pressure on the shoe. Foot pedals can be quite pronounced on some models and almost nonexistent on others.

A *bender hook* is the part of the bender shoe that holds the conduit in place during the bending process. The hook is used to determine the front of the bender and indicates which direction to point the bender. This distinction is critical when discussing three-bend saddles. Generally, the hook is the part of a bender that can be damaged, rendering the bender useless.

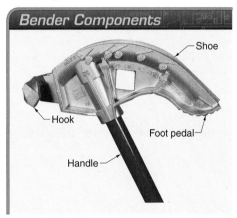

**Bender Components**

Shoe

Hook

Foot pedal

Handle

*Figure 2-4. The primary bender components are the shoe, foot pedal, handle, and hook.*

**Torpedo Levels**

*Figure 2-3. A torpedo level can be used to check the angles of several types of bends.*

**Bender Shoe Markings.** All benders have markings on the shoe. The arrow marking is the most used benchmark on any bender. **See Figure 2-5.** The arrow marking is used to make 90° bends and also as a benchmark on many other types of bends.

The star marking indicates the back of a 90° bend. Many benders also have rim notches located inside or outside of the shoe. These markings, which often look like file marks or teardrops, indicate the center of a 45° bend.

Angle markings vary greatly between manufacturers and in practical use. Some benders have built-in bubble levels that indicate 45° and 90° bends. Others use a pin and a graduated scale device that allow the operator to sight down the handle to determine the angle.

Some benders employ a series of markings on the outside of the shoe. When the conduit is parallel to a marking, that particular angle of bend has been achieved. Other benders use a small plumb bob to indicate the angle. Finally, some benders have no angle marks at all. Regardless of the type of bender used, an initial check should be made to determine the accuracy of the angle markings.

When bending conduit, the conduit is formed in the shape of the shoe. Keeping the shoe in good shape is critical to the quality of the final product. Using a conduit bender to bend other products, such as re-bar, or as a pry bar or lever will significantly reduce its life.

## Hickeys

A *hickey* is a hand bender with a no-radius shoe. **See Figure 2-6.** Hickeys are effective only on heavy-walled rigid conduit. If hickeys are used on EMT, the wall will collapse and the pipe will kink. Hickeys are available in ½″, ¾″, and 1″ rigid pipe sizes.

**Hickey Styles**

*Figure 2-6. A hickey is a hand bender with a no-radius shoe.*

It is very difficult to make accurate repeatable bends with hickeys. However, they can be used in slab installations where radius is often irrelevant. They are also quite handy when making adjustments to conduits coming out of a slab.

The technique for using a hickey is quite different than for a hand bender with a shoe. Foot pressure is not used and the bends are all accomplished with upper body strength. A bend with a hickey is fabricated by making several short bends at different locations along the conduit. **See Figure 2-7.** If a tight radius is desired, the bends are made close to one another. If a long radius is desired, the distance between the bends is increased.

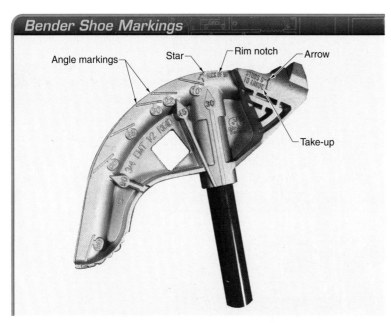

**Bender Shoe Markings**

Angle markings — Star — Rim notch — Arrow — Take-up

*Figure 2-5. Common markings on a bender shoe include the angle markings, star, rim notch, and arrow. In addition, the take-up is often given on the bender.*

## Conduit Reamers

A *conduit reamer* is a tool used to remove burrs and sharp edges from a piece of conduit after it has been cut to length. **See Figure 2-8.** Sharp edges can damage wire insulation and cause electrical shorts. A reamer is inserted into the cut end of the conduit and turned. The hooked edge of the reamer cleans up any roughness or burrs on the cut end. Burrs and sharp edges must be removed before conduit is installed.

## Tape Measures and Rules

A *tape measure* is a measuring device with a metallic tape wound up in a coil that can be extended to take measurements. A *rule* is a rigid measuring tool. **See Figure 2-9.** Tape measures and rules are used to measure distances and lengths in many conduit-bending applications.

A tape measure is a fundamental tool for construction personnel. Reading a tape measure is a critical skill that every electrician must have. A tape measure has markings representing distances in feet, inches, and fractional parts of an inch.

The hook end of a tape is intentionally loose. It moves back-and-forth a distance equal to its thickness. This allows measurements to be made by pulling the tape, with the hook holding the end secure, or by pushing the tape against an edge. The hook end slides and allows a correct measurement to be made in either situation.

**Hickey in Use**

**Figure 2-7.** *A bend with a hickey is fabricated by making several short bends at different locations along the conduit.*

**EMT Reaming Tools**

**CONDUIT REAMER**   **CONDUIT REAMING SCREWDRIVER**

*Klein Tools, Inc.*

**Figure 2-8.** *A conduit reamer is a tool used to remove sharp edges from a piece of EMT after it has been cut to length.*

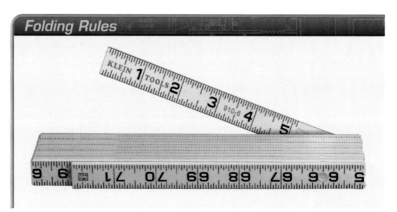

**Folding Rules**

*Klein Tools, Inc.*

**Figure 2-9.** *A folding rule is a measuring device that folds up for easy handling.*

---

**Tech Fact**

Folding rules come in several varieties. The inside reading variety is the type most commonly used in electrical work. Tape measures also come in several varieties. In every case, the hook end must be protected. If the hook becomes bent, such as from a fall, any readings made using the hook will be in error. In addition, the only practical tape measures for fieldwork are 1″ or more in width. Narrower tapes are not rigid enough to be extended for a measurement.

**Converting between Fractions and Decimals.** It is very useful to be able to convert between fractions of an inch and their decimal equivalents. Many calculations involving fractions are done on calculators. The fractions must be converted to decimal numbers to enter into the calculator. After the calculations are completed, the decimal number needs to be converted back to fractions in order to use a tape measure.

The simplest method for converting between decimals and fractions is to use a conversion table. **See Figure 2-10.** To convert from a fraction to a decimal, the fraction is located in the table and the equivalent decimal is found by following the line across in the table. To convert from a decimal to a fraction, the decimal number is found and the line followed back to its equivalent fraction.

For example, from the table the fraction $\frac{5}{16}$ is equal to 0.3125. Similarly, if the result of a calculation is 0.625, this is equal to the fraction $\frac{5}{8}$. If the decimal number is not in the table, select the value in the table that is nearest to a decimal. If the result of a calculation is 0.6, the nearest decimal in the table is 0.625 and the equivalent fraction is $\frac{5}{8}$.

If a calculator is available, any fraction can be converted to a decimal by dividing the numbers. For example, the fraction $\frac{3}{4}$ can be converted to decimal by dividing the 3 by the 4, resulting in 0.75. Similarly, the fraction $\frac{5}{16}$ can be converted to a decimal by dividing the 5 by the 16, resulting in 0.3125.

A calculator can also be used to convert a decimal number to a fraction. Simply multiply the decimal number by 16. This gives the number of 16ths in the fraction. For example, the decimal number 0.62 multiplied by 16 gives 9.92, or approximately 10. Therefore, the decimal number 0.62 is approximately equal to $\frac{10}{16}$, or $\frac{5}{8}$. **See Figure 2-11.**

For mixed numbers, care must be taken to multiply only the decimal part by 16. Multiplying the entire mixed number by 16 gives the wrong results. Only the part to the right of the decimal point is multiplied by 16. If a calculation result of 20.80″ is multiplied by 16, the result is 332.8, which has no meaning. If the part to the right of the decimal point, 0.80, is multiplied by 16, the result is 12.8, or about 13. This means that 20.80″ is approximately equal to $20\frac{13}{16}″$.

| Fraction and Decimal Conversions | |
|---|---|
| Fraction | Decimal |
| $\frac{1}{16}$ | 0.0625 |
| $\frac{2}{16}$, $\frac{1}{8}$ | 0.125 |
| $\frac{3}{16}$ | 0.1875 |
| $\frac{4}{16}$, $\frac{2}{8}$, $\frac{1}{4}$ | 0.25 |
| $\frac{5}{16}$ | 0.3125 |
| $\frac{6}{16}$, $\frac{3}{8}$ | 0.375 |
| $\frac{7}{16}$ | 0.4375 |
| $\frac{8}{16}$, $\frac{4}{8}$, $\frac{2}{4}$, $\frac{1}{2}$ | 0.5 |
| $\frac{9}{16}$ | 0.5625 |
| $\frac{10}{16}$, $\frac{5}{8}$ | 0.625 |
| $\frac{11}{16}$ | 0.6875 |
| $\frac{12}{16}$, $\frac{6}{8}$, $\frac{3}{4}$ | 0.75 |
| $\frac{13}{16}$ | 0.8125 |
| $\frac{14}{16}$, $\frac{7}{8}$ | 0.875 |
| $\frac{15}{16}$ | 0.9375 |
| $\frac{16}{16}$ | 1 |

*Figure 2-10. Some fractions can be represented in more than one way, and all fractions have an equivalent decimal value.*

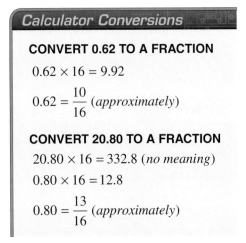

**Calculator Conversions**

**CONVERT 0.62 TO A FRACTION**

$0.62 \times 16 = 9.92$

$0.62 = \frac{10}{16}$ *(approximately)*

**CONVERT 20.80 TO A FRACTION**

$20.80 \times 16 = 332.8$ *(no meaning)*

$0.80 \times 16 = 12.8$

$0.80 = \frac{13}{16}$ *(approximately)*

*Figure 2-11. A calculator can be used to convert a decimal number to a fraction.*

## Mental Math

If a calculator or conversion table is not available, there are other simple methods that can be used to convert between decimals and fractions. These methods are not exact, but give results to the nearest 16th of an inch.

In the first method, the decimal value can be expressed in hundredths and divided by 6. The result is a close estimate of the number of 16ths. For example, the decimal 0.80 is $^{80}/_{100}$, or 80 hundredths. Dividing the 80 by 6 results in 13.3. Rounding 13.3 gives 13, indicating that 0.80 is approximately equal to $^{13}/_{16}$. For example, a calculation result of 20.80″ is approximately equal to 20$^{13}/_{16}$″.

The second method requires memorizing the decimal equivalents for all of the $^{1}/_{8}$″ values in their 100ths form and then simply adding or subtracting 6 to come up with the nearest corresponding $^{1}/_{16}$″ equivalent. For example, $^{3}/_{8}$″ (0.375″) is approximately equal to 0.38, or 38 hundredths. Therefore, the value of the next highest 16th, $^{7}/_{16}$, can be determined by adding 6 to the 38. The sum of 38 and 6 is 44, indicating that $^{7}/_{16}$ is fairly close to 0.44. The actual value is 0.4375.

The value of the next lowest 16th, $^{5}/_{16}$, can be determined by subtracting 6 from the 38. The difference between the 38 and 6 is 32, indicating that $^{5}/_{16}$ is fairly close to 0.32. The actual value is 0.3125.

### Conversions without Calculator

$$0.80 = \frac{80}{100}$$

$$\frac{80}{6} = 13.3$$

$$0.80 = \frac{13}{16}$$
*(approximately)*

**DIVIDING BY 6**

| Eighths | Decimal | Hundredths |
|---------|---------|------------|
| $^{1}/_{8}$ | 0.125 | 13 |
| $^{2}/_{8}$ | 0.25 | 25 |
| $^{3}/_{8}$ | 0.375 | 38 |
| $^{4}/_{8}$ | 0.5 | 50 |
| $^{5}/_{8}$ | 0.625 | 63 |
| $^{6}/_{8}$ | 0.75 | 75 |
| $^{7}/_{8}$ | 0.875 | 88 |
| $^{8}/_{8}$ | 1 | 100 |

$$\frac{3}{8} = 38 \ (hundredths)$$

$$\frac{7}{16} = 38 + 6 = 44 \ (hundredths)$$

$$\frac{7}{16} = 0.44 \ (approximately)$$

$$\frac{5}{16} = 38 - 6 = 32 \ (hundredths)$$

$$\frac{5}{16} = 0.32 \ (approximately)$$

**ADDING AND SUBTRACTING 6**

## BASIC BENDS

A 90° bend is the first bend learned by an electrician. The techniques used in making this basic bend are used in many other types of bends. Take-up is an adjustment made to a measurement when making bends.

### Tech Fact

The thin walls of EMT can easily kink and bend out of shape when bent with a hickey.

## Take-up

Every bender has a take-up, or deduction. *Take-up* is the value that is used to determine where to place the bending marks. The take-up is often stamped on hand benders. The take-up can also be determined from the bender manual or chart provided with a new bender. **See Figure 2-12.** An electrician should not assume that the take-up is the same for all benders of the same size. In addition, the take-up may vary from one manufacturer to another.

| Take-up | |
|---|---|
| Bender Size | Typical Take-up |
| ½″ | 5″ |
| ¾″ | 6″ |
| 1″ | 8″ |
| 1¼″ | 11″ |

*Figure 2-12. Every bender has a specific take-up. This value is used to determine where to place pencil marks on the conduit before it is bent.*

## 90° Bends

In order to place the bending mark in the correct position on the conduit, the take-up must first be subtracted from the desired stub length. A pencil mark is then placed on the conduit where the bend is to be made. A soft lead pencil is recommended so that the mark can be easily removed if necessary. Marks made with a permanent marker can be an unnecessary distraction on exposed conduit and are considered unprofessional. The pencil mark should be drawn entirely around the conduit. This makes it much easier to find the mark when making multiple bends in low lighting conditions, or when the conduit is rotated during the bending process.

A bend is made using heavy foot pressure. The handle is used only to guide the bend. If bends in EMT are made using the handle to apply pressure to the shoe, the conduit can rise up from the bending surface. This creates either a distorted curve in the bend or a kink in the conduit. When bending heavier-walled rigid conduit, foot pressure is just as critical; however, some force will need to be applied with the handle to complete the bend.

**Making Stub-up Bends.** A *stub-up bend* is a 90° bend in conduit made perpendicular to the original length of the conduit, with the conduit extending a specified length from the back of the bend. **See Figure 2-13.** A stub-up bend can be used whenever conduit has to make a 90° turn. However, a stub-up bend still requires a

simple calculation to determine where to place the bend mark in order to reach a desired stub length.

A stub-up bend is created as follows:

1. Measure the length of the required stub-up. This is the distance from the back edge of the conduit out to where the conduit should end. **See Figure 2-14.**

2. Subtract the take-up from the stub length measurement. Place a pencil mark on the conduit at the calculated distance.

3. Place the bender on the conduit and align the arrow benchmark on the bender with the pencil mark on the conduit. Make certain the bender hook is facing toward the end of the conduit from which the measurement was taken.

4. Place the conduit and bender on a hard level surface. Place one foot on the bender foot pedal and grasp the handle of the bender with both hands. The other foot may remain on the ground or may be placed on the conduit to help steady it. Use the handle to guide the bender while using heavy foot pressure to make the bend.

5. Once the bend approaches 90°, check your progress with a torpedo level, making sure the surface on which you are working is level. Some fine-tuning is often necessary to make a perfect 90° bend. Bending slightly past 90° is usually required to compensate for springback.

6. Check the completed stub length against the desired length. It is often a result of poor foot pressure if the completed stub length is longer than the desired length.

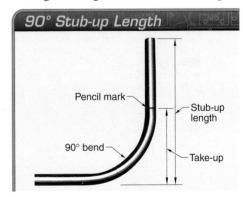

*Figure 2-13. A stub-up bend is fabricated to a precise stub-up length.*

## 90° Stub-up Bending

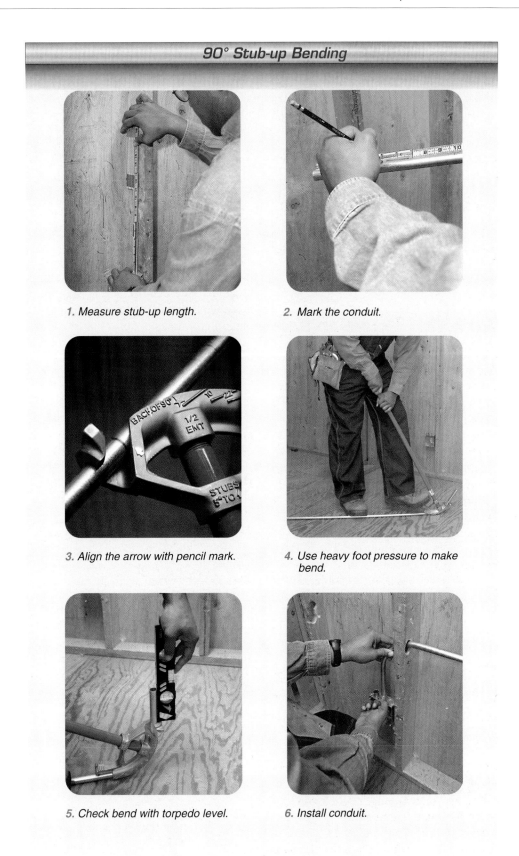

1. Measure stub-up length.

2. Mark the conduit.

3. Align the arrow with pencil mark.

4. Use heavy foot pressure to make bend.

5. Check bend with torpedo level.

6. Install conduit.

*Figure 2-14.* A stub-up bend is started at the pencil mark.

For example, a ½″ EMT bender with a 5″ take-up is used to make a stub-up bend. **See Figure 2-15.** In the first step, the length of the stub is measured at 10″. Next, the take-up of 5″ is subtracted from the 10″ length of the stub-up, which results in 5″. A pencil mark is placed on the conduit at a distance of 5″ from the end.

The bender is placed on the conduit and the arrow is aligned with the pencil mark. After the bender is placed on the conduit, heavy foot pressure is used to make the bend. The finished bend is checked with a torpedo level and the overall length is checked against the desired length.

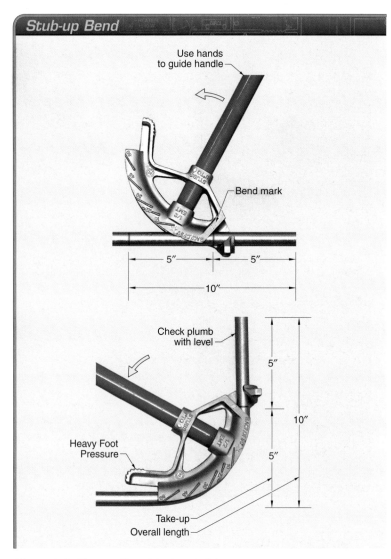

**Figure 2-15.** The pencil mark is placed 5″ from the end when the stub-up length is 10″ and the take-up is 5″.

**Measuring Shortcuts.** There are two simple shortcuts that can be used when working with take-up to eliminate the possibility of the arithmetic errors. **See Figure 2-16.**

The first shortcut is to place the measuring tape with the take-up length at the end of the conduit and place the pencil mark at the tape marking for the stub-up length. For example, if the measured stub-up length is 15″ and the take-up is 7″, the end of the conduit is placed at the 7″ marking and the pencil mark is placed at the 15″ marking.

The second shortcut consists of positioning a measuring tape with the stub-up length measurement at the end of the conduit and placing the pencil mark at the tape marking for the take-up.

For example, if the measured stub-up length is 15″ and the take-up is 7″, the end of the conduit is placed at the 15″ marking on the tape. The pencil mark is placed on the conduit at the 7″ marking.

**Determining an Unknown Take-up.** There are situations where the take-up for a bender is not known. In this case, an electrician can determine the actual take-up of a bender by using a scrap piece of conduit. The actual take-up measurement is found by placing a pencil mark on the scrap conduit at any convenient location and making a 90° bend with the arrow at that mark.

The distance from the back of the bend to the pencil mark is the take-up. The distance is measured by laying the conduit flat on the ground and placing a straight edge against the back of the bend and measuring to the pencil mark. This is a universal method of finding take-up. It works for all types of benders and shoes.

## Bend Corrections

There are occasions when an error has been made in bending conduit. When a bend ends up being less than the required 90°, the bend can be fixed by placing the bender back on the conduit at the same benchmark and bending slightly more. If the bend ends up being bent past 90°, the bend can be fixed by placing the handle of the bender over the stub gently bending it backward. This requires that the handle be placed as far down the stub as possible to avoid a kink in the bend. One foot should be placed on the conduit to steady it during the straightening.

### Back-to-Back 90° Bends

Fabricating back-to-back 90° bends is the next skill to be mastered. Back-to-back 90° bends consist of two 90° bends fabricated on the same length of conduit. The bends may be required to be in the same direction, 90° to the left or right, or in the opposite direction. Back-to-back 90° bends can be used to fit conduit within a space between obstructions or to connect two boxes a known distance apart.

**Making Back-to-Back Bends in the Same Direction.** There are three measurements for any back-to-back 90° bend. **See Figure 2-17.** The first two measurements are the lengths of the two stubs. The third measurement is the back-to-back distance (distance from the back of one bend to the back of the other). The back-to-back bend is fabricated as follows:

1. Measure the two stub-ups and the required back-to-back distance. **See Figure 2-18.**

2. Use the first measured stub-up length and the bender take-up to make the first bend. This is the same procedure as discussed previously.

3. Subtract the bender take-up from the measured back-to-back distance. This difference is the distance from the back of the first bend to the pencil mark for the second bend. Place a straightedge along the first stub to extend the back of the first 90° bend. Measure from the straightedge away from the first bend and place a pencil mark at the calculated distance.

4. Place the bender on the conduit and align the arrow with the pencil mark. It is very important to keep the two bends in the same plane. Carefully sight along the bender and conduit to make sure that the second bend is aligned with the first. Make certain the bender hook is pointing toward the first bend and make the second bend.

5. Check the bend with a level. Measure the length of the second stub-up and place a pencil mark on the conduit. The stub length is measured from the ground or from a straightedge placed on the back of the second bend.

6. Once the bends are complete, the conduit can be cut to length, reamed, and installed.

#### Tech Fact

A dogleg is an undesirable bend that is not in line with other bends in the same conduit. A conduit with a dogleg does not lie flat against the wall or ceiling. A dogleg is sometimes called a "wow".

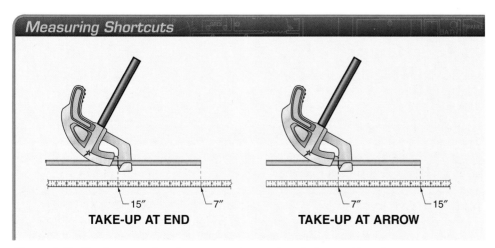

**Measuring Shortcuts**

15″  7″
**TAKE-UP AT END**

7″  15″
**TAKE-UP AT ARROW**

*Figure 2-16. Simple measuring shortcuts can be used to eliminate arithmetic errors.*

## Reverse Method for 90° Bends

Another method for laying out and forming a 90° bend is called the reverse method or the B method. When using this method to form a 90° bend, the take-up for the bending shoe is no longer a consideration. The reverse method is often used when forming 90° bends with long leg lengths. Forming 90° bends with long leg lengths can be accomplished with the standard take-up method; however, it can be awkward or difficult to control the conduit in the bender during the bending process.

The reverse method forms the short leg as if it were a stub while leaving the long stub on the ground as if it were a leg. For example, if the desired stub of a 90° bend is 95″ long, the leg remaining on the floor is only 25″ long plus the gain. A stub of this length is difficult to handle, so the bend can be made in reverse. When this bend is complete, a leg length is formed that is 95″ long from the end of the conduit to the back of the 90° bend. The reverse method can be used to form a 90° bend as follows:

1. Measure 95″ from one end of the conduit and place a pencil mark.

2. Place the bender on the conduit with the hook pointing towards the opposite end of the conduit (the short end). Align the pencil mark and the point of the star.

3. Fabricate the 90° bend, bending the short end in the air.

The reverse method of fabricating 90° bends can also be used in concrete slab or pan applications. This method can be used when the conduit is in place and the partition is marked on the deck. The bend is made by aligning the star with the partition mark.

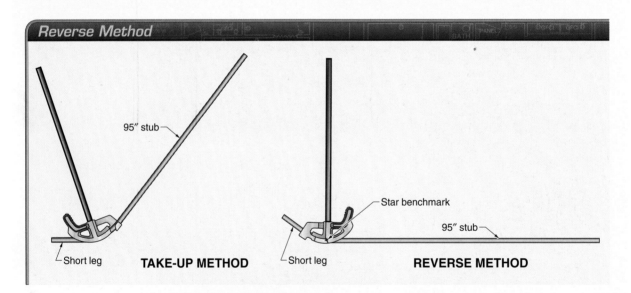

**Reverse Method**

95″ stub

Short leg — **TAKE-UP METHOD**

Star benchmark

95″ stub

Short leg — **REVERSE METHOD**

**Making Bends at Right Angles.** If the situation calls for the two 90° bends to be at right angles to each other, the procedure is very similar to that for back-to-back bends in the same direction. This situation may occur when conduit is run along one wall, makes an elevation change at a corner (the first 90° bend), and then makes another 90° bend to run along another wall.

The only difference between this procedure and making a standard back-to-back bend is the angle of the second bend rela-tive to the first bend. **See Figure 2-19.** For best accuracy, the bender is placed on the conduit and the conduit rotated until a level shows that the first bend is parallel to the floor. The arrow benchmark and the pencil mark are aligned and the second bend is then completed in the normal manner.

**Tech Fact**

Take-up is the distance measured from the arrow on the bender to the back side of the conduit after making a 90° bend.

## BEND PRE-POSITIONING

There are many situations where it is desirable to measure and cut the conduit to the proper length before fabricating a 90° bend. It is far easier to thread a straight length of conduit than conduit that has been bent. The principle of gain allows a section of conduit to be cut to the proper length before bending.

### Gain

Because conduit is required to be bent with a radius, the actual length of the conduit will be less than the sum of the horizontal and vertical distances as measured in the run. *Gain* is the difference between the sum of the straight distances and the actual length of conduit. **See Figure 2-20.** In other words, gain is the "shortcut" provided by following the arc of the curve instead of the straight distance.

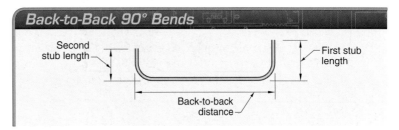

**Figure 2-17.** *A back-to-back bend requires three measurements.*

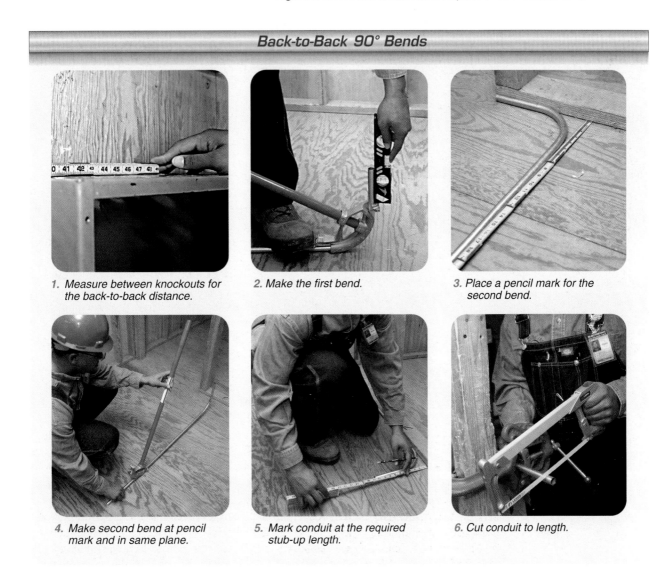

1. Measure between knockouts for the back-to-back distance.

2. Make the first bend.

3. Place a pencil mark for the second bend.

4. Make second bend at pencil mark and in same plane.

5. Mark conduit at the required stub-up length.

6. Cut conduit to length.

**Figure 2-18.** *When back-to-back bends are made in the same direction, the bend is fabricated on the floor.*

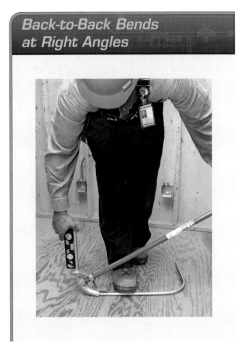

*Figure 2-19. Back-to-back bends are often made at right angles to each other.*

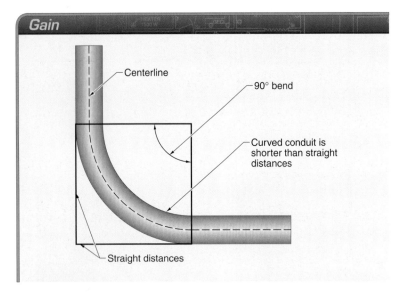

*Figure 2-20. Gain is the shortcut provided by following the arc of a bend instead of the straight distances.*

**Tech Fact**

A larger bend radius allows the wire to be pulled more easily than a smaller bend radius. However, a smaller bend radius allows the bend to fit tighter in a corner.

To find the length of straight conduit needed to complete a run, the straight distances are added and the gain is subtracted from this sum. The difference is the linear length of conduit required. The amount of gain for conduit of different sizes may be available from the bender manufacturer. **See Figure 2-21.**

For example, a run of ¾″ EMT has a leg of 37½″ and a stub of 33½″. The gain from the table is 3″. **See Figure 2-22.** The needed length of conduit before making the bend is 68″ (37½ + 33½ – 3 = 68).

**Measuring Gain.** Generally speaking, there is no standard value for typical gain. Gain should be determined the first time a particular bender is used. In this situation, gain can easily be measured by using a scrap piece of conduit. Gain can be measured as follows:

1. Measure the length of a scrap piece of conduit.

2. Fabricate a 90° bend and verify that the bend is accurate.

3. Measure the stub and leg lengths.

The sum of the stub and leg lengths is the completed length of the 90° bend in the conduit. The difference between the completed length and the original length is the gain for that bender.

For example, ¾″ EMT is being used on a job. A scrap piece of conduit measuring 36⅝″ long is used to determine gain. After fabricating a 90° bend, the stub is 15⅝″ long and the leg is 24⅛″ long. The completed length is 39¾″ (15⅝ + 24 ⅛ = 39¾). Since the gain is the difference between the completed length and the original length, the gain is 3⅛″ (39¾ – 36⅝ = 3⅛).

**Back-to-Back Gain.** The principle of gain can be used when bending back-to-back 90° bends. Each of the two bends has an equal amount of gain, so the total amount of gain is twice the gain of a single bend. To find the total precut length of conduit for a back-to-back installation, simply add the length of the two 90° stubs to the distance between the backs of the 90° bends and subtract twice the gain. If the run has three 90° bends, subtract three times the gain. The required length of conduit for a two-bend back-to-back 90° bend is calculated as follows:

$$l = S_1 + S_2 + D - (gain \times 2)$$

where

$l$ = precut length, in inches
$S_1$ = length of stub-up 1, in inches
$S_2$ = length of stub-up 2, in inches
$D$ = distance between backs of the bends, in inches
$gain$ = gain, in inches

For example, a back-to-back 90° bend is made in ½″ EMT. **See Figure 2-23.** The length of one stub-up is 13¾″ and the length of the other stub-up is 22¼″. The distance between the backs of the bends is 41″. For ½″ EMT, the gain is 2½″. The required length of conduit is calculated as follows:

$$l = S_1 + S_2 + D - (gain \times 2)$$

$$l = 13\tfrac{3}{4} + 22\tfrac{1}{4} + 41 - (2\tfrac{1}{2} \times 2)$$

$$l = 77 - 5$$

$$l = \mathbf{72''}$$

## APPLICATION—THREE-BEND BACK-TO-BACK BENDS

An electrician needs to bend conduit around several obstacles and decides to use a three-bend back-to-back 90° bend. **See Figure 2-24.** The conduit is ½″ rigid bent on a ¾″ EMT bender with a gain of 3⅟16″ and a take-up of 6″. The two stubs are 12″ and 16″, with back-to-back distances of 21″ and 36″, with the stubs at a right angle to the legs.

### Tech Fact

Threadless couplings and connectors used with conduit shall be made tight. Where buried in masonry or concrete, they shall be concrete tight. Threadless couplings and connectors shall not be used on threaded conduit ends unless listed for the purpose. Running threads shall not be used on conduit for connection at couplings.

**Hand Bending Gain**

| Conduit Size | Typical EMT Bend Radius | EMT Gain | Typical Rigid Bend Radius | Rigid Gain |
|---|---|---|---|---|
| ½″ | 4³⁄16″ | 2½″ | 5⅛″ | 3⅟16″ |
| ¾″ | 5⅛″ | 3″ | 6½″ | 3¹³⁄16″ |
| 1″ | 6½″ | 3¹⁵⁄16″ | 9⅝″ | 5⁷⁄16″ |
| 1¼″ | 8″ | 4¹⁵⁄16″ | – | – |
| | 9⅝″ | 5⅝″ | – | – |

***Figure 2-21.*** *The amount of gain depends on the conduit size and the bend radius. These values are approximate and should be verified for each bender.*

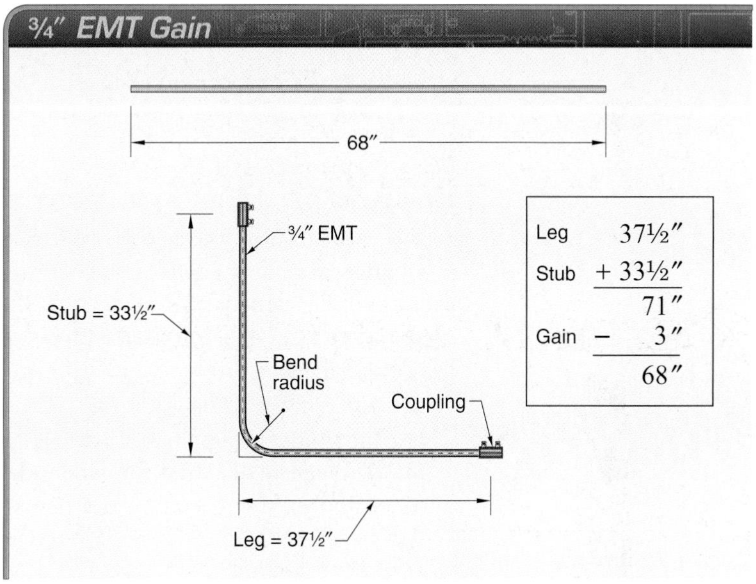
¾″ EMT Gain
68″
¾″ EMT
Stub = 33½″
Bend radius
Coupling
Leg = 37½″
Leg 37½″
Stub + 33½″
71″
Gain − 3″
68″

***Figure 2-22.*** *The conduit length is calculated from the horizontal and vertical distances and the gain.*

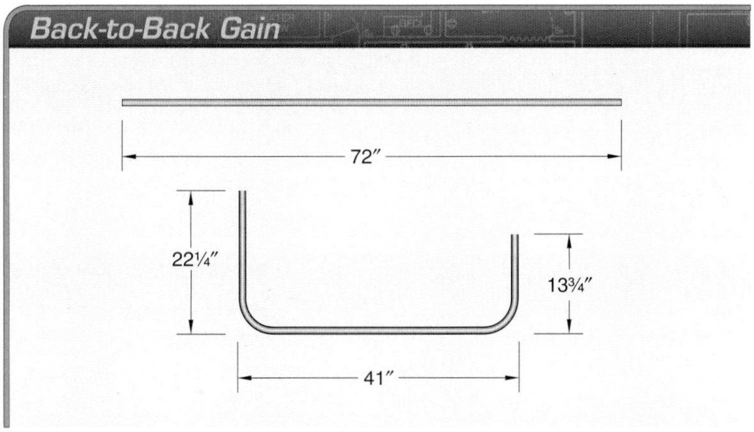
Back-to-Back Gain
72″
22¼″
13¾″
41″

***Figure 2-23.*** *The gain is doubled in the calculation for a back-to-back bend.*

### Calculating Conduit Length

The method used to calculate prebend length needs to be modified slightly for a three-bend back-to-back gain calculation. The total back-to-back distance needs to include both the legs and the gain needs to be multiplied by three because there are three bends. The required length of conduit is calculated as follows:

$$l = S_1 + S_2 + D_1 + D_2 - (gain \times 3)$$
$$l = 12 + 16 + 21 + 36 - (3\tfrac{1}{16} \times 3)$$
$$l = 85 - 9\tfrac{3}{16}$$
$$\boldsymbol{l = 75\tfrac{13}{16}}$$

The total required conduit length is 75¹³⁄₁₆″. Therefore, a 10′ length needs to be cut down to 75¹³⁄₁₆″ and threaded.

### Back-to-Back Bend Layout and Fabrication

A three-bend back-to-back bend is laid out in a manner similar to a two-bend back-to-back bend. The bender take-up needs to be used to determine the correct location for the pencil marks. A three-bend back-to-back bend is fabricated as follows:

1. Use the bender take-up of 6″ to calculate the distance to the first pencil mark. Since the first stub is 12″, the pencil mark is placed 6″ (12 – 6 = 6) from one end of the conduit.

2. Place the bender on the conduit and align the arrow benchmark with the pencil mark and make a 90° bend.

3. Use the bender take-up to calculate the distance to the second pencil mark. Since the first leg is 21″, the pencil mark is 15″ (21 – 6 = 15) from the back of the bend. Use a straightedge to place a pencil mark 15″ from the back of the bend.

4. Place the bender on the conduit and align the arrow benchmark with the new pencil mark and make a 90° bend. Be very careful to make sure that the bends are in the correct plane so a dogleg is not created.

5. For the third bend, align the first leg (21″) with a straight edge and measure 36″ from the back of the second bend. Place a pencil mark at that point.

6. Turn the bender around, facing the far end of the conduit (16″ stub), and align the star benchmark on the bender with the pencil mark on the conduit. This is the reverse method. Pull up on the 16″ stub while using the previous bend to stabilize the conduit on the floor. By keeping the 21″ leg flat on the floor it will be easier to finish the 16″ stub and have it be perpendicular to the previous bend.

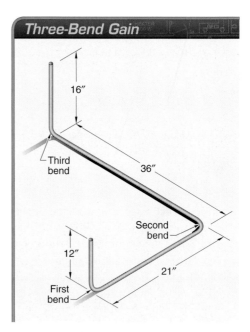

**Three-Bend Gain**

16″

Third bend

36″

12″

Second bend

First bend

21″

*Figure 2-24. The gain is multiplied by three in the calculation for a three-bend back-to-back bend.*

## Back-to-Back Bend Pre-positioning

Gain can be used to pre-position back-to-back 90° bends. This eliminates the need to measure the back-to-back distance while making bends. The gain for different types of conduit is available in tables. For ½″ rigid conduit bent on an EMT bender, the take-up is 6″ and the gain is 3⅟16″. For example, a three-bend back-to-back bend is needed with a 12″ stub, a 21″ leg, a 36″ leg, and a 16″ stub. The bends are fabricated as follows:

1. Use the bender take-up of 6″ to calculate the distance to the first pencil mark. Since the first stub is 12″, place the first pencil mark 6″ (12 − 6 = 6) from one end of the conduit.

2. The first leg is 21″. Taking the gain into account, place the second pencil mark 17⅟16″ (21 − 3⅟16 = 17⅟16) from the first pencil mark.

3. The second leg is 36″. Taking the gain into account, place the third pencil mark 32⅟16″ (36 − 3⅟16 = 32⅟16) from the second pencil mark.

4. Align the arrow benchmark with the first pencil mark and fabricate the 12″ stub.

5. Align the arrow benchmark with the second pencil mark and fabricate the 21″ leg.

6. Align the arrow benchmark with the third pencil mark and fabricate the 36″ leg.

The bender must face the starting end (12″ stub) for all three bends. The third bend may be difficult to fabricate unless the bender is turned around. This problem can be solved by simply marking the far end of the conduit at 10″ (16 − 6 = 10), reversing the direction of the bender, and bending the 16″ stub using the arrow benchmark.

## SUMMARY

- Standard hand benders are available from several manufacturers and come in sizes that bend ½″ to 1¼″ EMT conduit.

- Rigid conduit may also be bent with EMT hand benders. However, the next larger size bender must be used.

- A bender shoe is the curved part of a bender that holds the conduit during fabrication.

- A bender foot pedal is the part of the bender where foot pressure is applied in order to bend the conduit.

- A bender hook is the part of the bender shoe that holds the conduit in place during the bending process.

- A bender take-up, or deduction, is a value that is used to determine where to place the bending marks when making a stub-up or 90° bend.

- Back-to-back 90° bends consist of two or more 90° bends fabricated on the same length of conduit.

- Gain is the difference between the sum of the straight distances and the actual length of conduit.

# Hand Bending — Offsets and Kicks

## CONDUIT BENDING and FABRICATION

**3**

OFFSET BENDS ................................................................................. 30
KICKS ............................................................................................... 52
APPLICATION – PARALLEL OFFSETS WITH CONDUIT OF DIFFERENT SIZES ...... 54
SUMMARY ......................................................................................... 56

After 90° bends, offset bends are the most common bend. Offset bends may be used to change elevation, to get around an obstacle such as a structural member, another conduit, or ductwork, or to enter a junction box or an enclosure.

An understanding of basic trigonometry is useful in conduit bending. There are several angles that are commonly used in bending offsets. The most common angles for offset bends are 22½°, 30°, and 45°.

When an offset needs to be bent in a section of conduit, a right triangle can be drawn to represent the bend. The triangle represents the angle of the bend and the amount of offset. The triangle allows the use of trigonometry to find the distance between the centers of the two bends.

There are many situations where the bends must fall in a specific location on the conduit. This is often the case when conduit must be worked around an obstruction or when the bends of several conduits must be aligned to obtain a neat and workmanlike result. The total run length of conduit is reduced when an offset bend is made. This shrink must be taken into account when locating the bend points. Parallel offsets present unique challenges. Special techniques are used to keep the spacing even through the angled sections of the bends.

## OBJECTIVES

1. Define offset bend and kick and distinguish between them.
2. Demonstrate the multiplier method for making offset bends.
3. Calculate the distance between bends.
4. Describe shrink and explain how to calculate shrink.
5. Calculate the shift required when bending parallel offsets.
6. Demonstrate the measured rise method for making offset bends.
7. Demonstrate how to use shrink to calculate the required length of conduit before bending.

## OFFSET BENDS

After 90° bends, offset bends are the most common bend. An *offset bend* is a double conduit bend with two equal angles bent in opposite directions in the same plane in a conduit run. The bend may be used to get around an obstacle such as a structural member, another conduit, or ductwork or to enter a knockout in a box or an enclosure.

When making an offset bend, the electrician must take into account the location of the bend on the conduit, the angle chosen for the bend, and the amount of offset. The location of the bend is calculated based on the distance to the obstacle and the angle chosen for the bend. The angle for the bend is a compromise based on the space available for change and the ease of pulling wire.

The size of the obstacle determines the amount of offset required. For example, for conduit bent around a 4″ obstruction, the amount of the offset is the amount the conduit has to rise to clear the obstruction. **See Figure 3-1.** In this case, the conduit must rise 4″ to clear the obstruction.

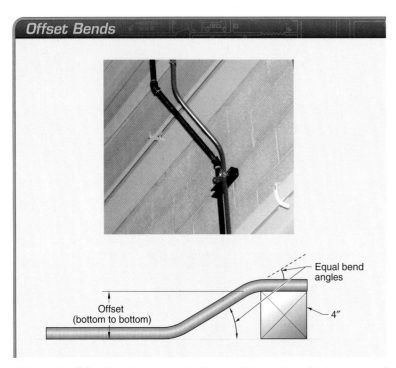

**Figure 3-1.** *Offset bends are used when conduit needs to be bent around obstacles.*

## Trigonometry of Offset Bends

An understanding of basic trigonometry is useful in conduit bending. *Trigonometry* is the branch of mathematics used to determine the sides and angles of triangles. The term "trigonometry" is often shortened to "trig" in common usage.

Commonly, standard angles are chosen for the bend angles in an offset. This allows the use of standard multipliers to calculate the distance between bends and the locations for the bends. Standard bend angles allow offsets to be reproduced and parallel offsets to be fabricated.

Electricians need to understand some trigonometry in order to understand how to make the simple calculations required for conduit bending. The study of trigonometry includes the study of angles, right triangles, and the trigonometry functions used to make calculations.

**Angles.** An *angle* is a measure of the rotation between two lines that are joined together at one point. **See Figure 3-2.** Angles are commonly measured in degrees and are designated with a degree symbol (°). The angle increases as the rotation between the lines increases.

There are several angles that are commonly used in bending offsets. These angles are 5° and 10°, commonly used for box offsets; 15°, 22½°, 30°, and 45° used for bending offsets and saddles; and 90°, which is used for bending stub-ups and other 90° bends. Other angles may be used, especially if the space for an offset is very tight.

**Right Triangles.** A *right angle* is an angle that measures exactly 90°. A *right triangle* is a triangle where one of the angles is a right angle. **See Figure 3-3.** When an offset needs to be bent in a section of conduit, a right triangle can be drawn to represent the bend. The triangle will represent the angle of the bend and the amount of offset. The triangle allows the use of trigonometry to find the distance between the centers of the two bends, provided the angle of bend is fabricated accurately.

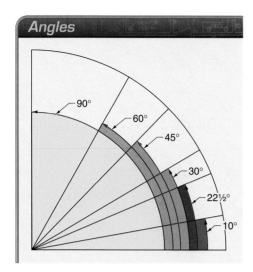

**Figure 3-2.** *An angle is a measure of the rotation between two lines that are joined together.*

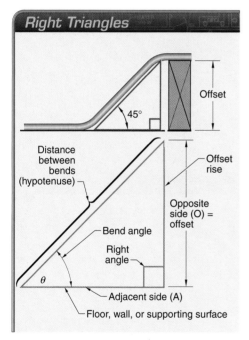

**Figure 3-3.** *A right triangle is a triangle where one of the angles is a right angle.*

When performing calculations with right triangles, one of the angles, other than the 90° angle, is chosen as the reference angle. The reference angle is the angle that the conduit is bent for the offset. The reference angle is often designated as angle theta ($\theta$). Theta ($\theta$) is a letter in the Greek alphabet often used by mathematicians to label angles.

The sides of the triangle are given names based on their location relative to the reference angle. The side of the triangle next to the reference angle and connecting to the right angle is called the adjacent side. The side of the triangle next to the reference angle and opposite the right angle is called the hypotenuse. The side opposite the reference angle is called the opposite side.

**Trigonometry Functions.** There are six ratios of the lengths of sides of right triangles. The ratios define the trigonometry functions. A function is a mathematical equation used to solve for unknown values. Six common trigonometry functions were created to describe the six ratios. These functions are the sine (sin), cosine (cos), tangent (tan), cosecant (csc), secant (sec), and cotangent (cot) functions. **See Figure 3-4.** These trigonometry functions can be used to calculate the lengths of the sides of the triangles for any bend angle.

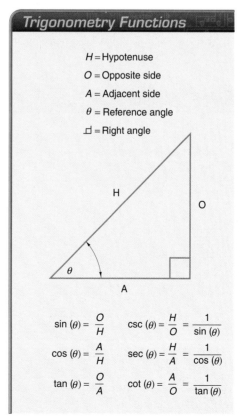

**Figure 3-4.** *There are six common trigonometry functions that use the ratios of the lengths of the sides.*

## Trigonometry Functions

The trigonometry functions are based on the ratios of the lengths of the sides of the triangle and are defined relative to the reference angle. For example, a function commonly used in conduit bending is the cosecant (csc) function. This function is conventionally written as follows:

$$csc(\theta) = \frac{hypotenuse}{opposite} \quad \text{or} \quad csc(\theta) = \frac{H}{O}$$

$$multiplier = \frac{distance\ between\ bends}{offset\ rise}$$

where

$\theta$ = bend angle
$csc(\theta)$ = cosecant of the bend angle
*hypotenuse* = length of the side opposite the right angle, in inches
*opposite* = length of the side opposite the bend angle, in inches
$H$ = hypotenuse
$O$ = opposite side

For any given bend angle, the ratios of the lengths of the sides of a right triangle are constant. For example, for a 30° bend angle in a triangle of any size, the ratio of the length of hypotenuse to the length of the opposite side is exactly 2:1. This is true for every 30° angle, regardless of the conduit size or the length of the opposite side (amount of offset).

For a 30° angle with an offset of 6″, the length of the hypotenuse of the triangle created by the bend is 12″. For a 30° angle with an offset of 8″, the length of the hypotenuse of the triangle created by the bend is 16″. In both of these cases, the cosecant of the reference angle is exactly 2. This is shown as follows:

$$csc(\theta) = \frac{H}{O}$$

$$csc(30°) = \frac{12}{6} = 2$$

$$csc(30°) = \frac{16}{8} = 2$$

The ratio of the lengths of the sides is different for each different reference angle. The cosecant of any angle can be looked up in a table or found with a scientific calculator by finding the sine of the angle and taking the reciprocal. The cosecant function is used in the multiplier method of bending offsets.

**Length Ratios**

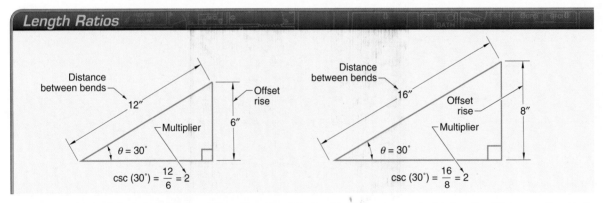

*For a 30° bend angle in a triangle of any size, the ratio of the length of hypotenuse to the length of the opposite side is exactly 2.*

### Multiplier Method for Offset Bends

The multiplier method for offset bends is very popular and can be used whenever a bender is available that can bend angles accurately. In order to make bends with the multiplier method, the offset rise must be measured, the bend angle chosen, the distance between bends calculated, and the two bends fabricated.

**Measuring the Offset Rise.** The first step in bending an offset is measuring the offset rise needed. The measurements must be taken from the same side of the conduit.

If the conduit is running tight against the floor, wall, or ceiling, the distance is measured from the bottom of the conduit at the wall to the bottom of the conduit where it crosses the obstacle. This measurement represents the distance from the bottom of the conduit against the wall to the bottom of the conduit against the obstacle. A measurement from the top of the conduit to the bottom of the conduit against the obstacle will be in error by the diameter of the conduit. **See Figure 3-5.**

If the conduit is some distance away from the wall, the offset rise is measured from the bottom of the conduit to where the conduit crosses the obstacle. The offset height is often called a rise, even if the measurement is down from the ceiling or out from the wall.

**Choosing the Bend Angle.** Once the offset rise is known, the next step is choosing the bend angle. A smaller bend angle makes it easier to pull wire through the conduit than a larger bend angle. However, a smaller bend angle takes up more room when rising over an obstacle than a larger bend angle. **See Figure 3-6.**

When different bend angles are used for the same offset, the run length and the distance between bends will be different. While both bends may work for a given installation, there are often field conditions that require the use of one particular bend angle over another. If the space for the bends is very tight, a 45° bend may be required. However, a smaller bend angle, such as a 30° bend, is generally preferred to make it easier to pull wire.

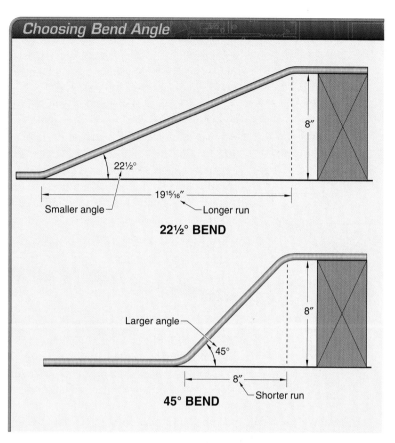

**Figure 3-6.** A smaller bend angle makes it easier to pull wire through the conduit than a larger bend angle. However, a smaller bend angle takes up more room when rising over an obstacle than a larger bend angle.

**Tech Fact**

All offsets are calculated on the basis of center-to-center measurements. Generally, offsets are measured bottom-to-bottom because it is the easiest way to take measurements. A bottom-to-top or top-to-bottom measurement will be in error.

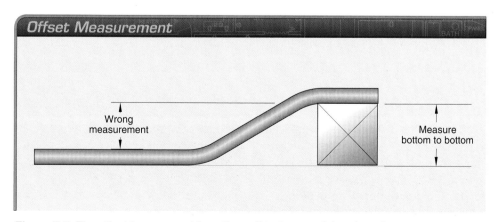

**Figure 3-5.** The offset is measured from the wall to the top of the obstacle.

The NEC® limits the total bends to the equivalent of 360° or less between wire pulling openings. In some cases, smaller bend angles may be needed to keep the total bend angles in compliance with the NEC®.

If the bend is not made carefully, the conduit will not be at the correct height to go over the obstacle. If the angle is bent too far (the bend angle is too large), the conduit will be too high as it crosses over the obstacle. If the angle is not bent far enough (the bend angle is too small), the conduit will not clear the obstacle.

**Calculating the Distance between Bends.** Once the bend angle is chosen, the next step is calculating the distance between bends. For any given bend angle, the ratio of the distance between the bends and the offset rise is constant. This ratio is the cosecant of the angle. For bends with standard angles, a simple table of distance multipliers is used. **See Figure 3-7.** In this table, the distance multiplier is equal to the cosecant of the bend angle.

The distance between bends is the product of the offset rise and the distance multiplier. Therefore, the distance between bends is calculated as follows:

$$Distance = multiplier \times offset\ rise$$

$$D = M \times O$$

where

$D$ = distance between bends, in inches
$O$ = offset rise, in inches
$M$ = distance multiplier

For example, the offset rise is 8¼″ (8.25″), and a 45° bend angle can be chosen. The multiplier for a 45° bend is 1.41. The distance between bends is calculated as follows:

$$D = M \times O$$

$$D = 1.41 \times 8.25$$

$$D = \mathbf{11.63''}$$

The calculated distance between bends is 11.63″, or 11⅝″. For bends with non-standard angles, a table of trigonometry values or a scientific calculator can be used to find the cosecant. **See Appendix.** The cosecant of the angle can be used as the distance multiplier.

| Distance Between Bends | |
|---|---|
| Bend Angle, θ | Distance Multiplier |
| 5° | 11.4 |
| 10° | 5.76 |
| 15° | 3.86 |
| 22½° | 2.61 |
| 30° | 2.00 |
| 45° | 1.41 |

**Figure 3-7.** *A table of multipliers for standard angles is helpful for calculating the distance between bends.*

**Marking the Bend Locations.** After the distance between bends is calculated, the next step is marking the bend locations. The procedure to mark the conduit for this offset bend is as follows:

1. Place a pencil mark on the conduit where the first bend is to be made. **See Figure 3-8.** The pencil mark should go completely around the conduit so it is visible from all angles.

2. Use the calculated distance between bends to measure from the first mark to the location of the second mark. For an offset bend with an 8¼″ offset and a 45° bend angle, the distance between bends is 11⅝″. In this case, place a pencil mark 11⅝″ from the first mark.

*Some offsets are used to move the location of a conduit run.*

## Multiplier Method for Offset Bends—Marking Bend Locations

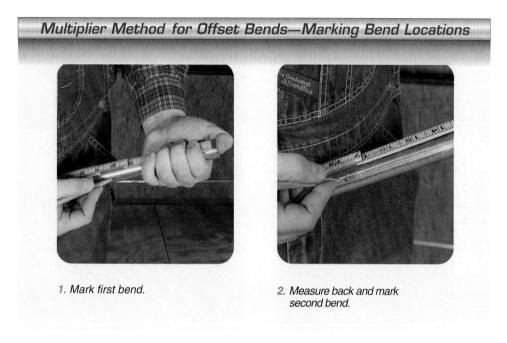

1. Mark first bend.

2. Measure back and mark second bend.

**Figure 3-8.** Pencil marks are placed on the conduit to indicate the location of the bends. The marks are separated by the calculated value of the distance between bends.

**Making the First Bend.** After the bend locations are marked on the conduit, the next step is making the first bend. The procedure to make the first bend in the conduit is as follows:

1. Align the first pencil mark with the arrow on the conduit bender and place the conduit and bender squarely on the floor. **See Figure 3-9.**

2. Use heavy foot pressure on the heel of the bender and pull up on the bender handle. Use the handle to guide the bend, not to make the bend. Maintain the foot pressure until the bend reaches the desired bend angle. Use the angle marks on the bender shoe to determine the correct amount of bend. There will be some springback so it will be necessary to bend slightly past the desired bend angle.

When learning to bend conduit, the bend angle should be checked with a protractor level to verify that it is correct. The level should be placed on the conduit, not the bender handle. With experience, this step should not be necessary.

The angle marks on a bender should be checked the first time the bender is used. A scrap conduit can be bent to the desired angle based on the markings on the bender. A protractor level or spirit level with the appropriate angle vial can be used to verify the angle of bend.

With practice, bends can regularly be made to within a degree or two every time the same bender is used. Because there are many types of benders and because of the slight differences between similar benders, an initial check should be made every time a new bender is used.

It is possible to use a protractor level on every single bend made on a job. However, this would drastically slow down the job. It is very important that an electrician learns to use a bender to produce consistent bend angles.

Step 1 says to align the pencil mark with the arrow on the bender shoe. This benchmark is convenient to use and easy to remember. However, it should be noted that any marking on the shoe can be used to make this bend as long as the same benchmark is used for both bends.

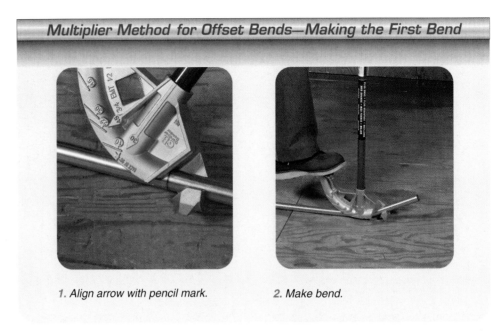

**Multiplier Method for Offset Bends—Making the First Bend**

*1. Align arrow with pencil mark.*

*2. Make bend.*

**Figure 3-9.** *The first bend is started by aligning the first pencil mark with the arrow on the bender shoe.*

**Making the Second Bend.** After the first bend is made, the second bend can be fabricated. To make the second bend in the correct direction, the bend must be made with the conduit held in the bender in the same direction as the first bend. The bender must not be removed from the conduit and reversed. The procedure to make the second bend in the conduit is as follows:

1. Turn the bender upside down, resting the handle solidly on the floor. **See Figure 3-10.** Move the conduit in the bender until the second mark is aligned with the arrow on the bender.

2. Rotate the conduit in the bender so that it is now turned 180° from its original position. Sight down the conduit and observe the first bend relative to the handle. The second bend must align with the first to keep the bends in the same plane. A good technique is to align the handle, the head of the bender, and the first bend with a straight edge, chalkline, or a straight line on the floor.

3. Start the second bend by applying hand pressure to the conduit as near to the bender shoe as possible. Do not apply pressure at a large distance from the bender shoe or the conduit may bend somewhere between the bender shoe and the point where the pressure is applied.

4. Make a small bend of about 5° to 10° and check the alignment of the two bends. Continue bending until the first bend disappears from view.

5. Stop bending before the bend reaches the full amount of bend. Turn the bender over so that the conduit is back on the floor. Make sure that the conduit does not move or rotate in the bender. Complete the bend on the floor.

6. Verify the accuracy of the offset by measuring the amount of rise actually created by the bends. Inspect the offset to verify that the bends were made to the correct angle and that a dogleg was not created.

---

### Tech Fact

The level of accuracy required for a given job is somewhat determined by the type of job. Concealed work does not require the same level of perfection that exposed work does and generally EMT doesn't require the same level of precision that a rigid metal conduit installation does. As a rule, the bends should come within 1/8″ of the desired value, but there may be times, especially in explosion-proof work, when 1/16″ accuracy is required.

## Multiplier Method for Offset Bends—Making the Second Bend

*1. Place bender upside down.*

*2. Align second bend with first.*

*3. Start second bend.*

*4. Make a small bend.*

*5. Turn bender over and finish bend.*

*6. Measure finished offset.*

**Figure 3-10.** *The second bend is started with the bender turned upside down.*

**Application—Offset around Duct.** A run of ¾″ EMT is being run along a ceiling. A rectangular ventilation duct is in the way of the run and an offset must be bent around the duct. **See Figure 3-11.** The amount of offset is measured at 9⅝″ (9.625″). A 45° angle is chosen for the bend. For a 45° angle, the distance multiplier is 1.41 and the distance between bends is calculated as follows:

$$D = O \times M$$
$$D = 9.625 \times 1.41$$
$$D = \textbf{13.57 or 13⁹⁄₁₆″}$$

The next step is to mark the bend locations. A pencil mark is made on the conduit where the first bend is to be made. Another pencil mark needs to be made 13⁹⁄₁₆″ from the first pencil mark. The first bend is made at the position of first mark. The arrow benchmark and the pencil mark need to be aligned.

The conduit and bender are placed on the floor and heavy foot pressure on the heel of the bender is used to make the 45° bend. After the first bend is complete, the bender needs to be turned upside down with the handle placed firmly on the floor.

For the second bend, the conduit is rotated 180° in the bender and the arrow benchmark aligned with the second pencil mark. After visually aligning the first bend with the handle, the second bend is started with hand and arm pressure. After the conduit has been bent enough that the first bend has disappeared from view, the bender is turned upside down with the conduit on the floor to complete the bend.

**Offset around Duct**

*Figure 3-11. An offset is used to bend conduit around an obstruction.*

### Bend Errors

There are several typical bend errors made when fabricating an offset. A *dogleg*, or *wow*, is a multiple bend in conduit where one of the bends is not in the same plane as the other bend. **See Figure 3-12.** A dogleg causes the offset to be skewed so it does not lie flat against the wall or ceiling.

Rotating and aligning the conduit for the second bend can be a challenge and requires some patience and practice in order to prevent a dogleg in the conduit. A dogleg occurs when the conduit has not been rotated exactly 180° from its position where the first bend was made.

The best way to avoid a dogleg is to visually inspect the alignment before making the second bend. Some electricians find it useful to pause and inspect the alignment after bending about 5° to 10° of the second bend. A little extra time spent on this step can save the time and labor used to straighten out a dogleg in a length of conduit.

Another bend error occurs when the two bends of the offset are not made at the same angle. When this happens, the top leg of the offset is not parallel with the bottom leg. This causes difficulties because the end of the conduit will not be in the correct location to attach the next conduit.

An *open offset bend* is an offset where the second bend is made with an angle that is too large and the end of the conduit rises relative to the top of the obstacle. A *closed offset bend* is an offset where the second bend is made with an angle that is too small and the end of the conduit falls relative to the top of the obstacle.

Another type of bend error occurs when the actual offset bend rises the wrong distance for the obstacle. If the bends are made with angles that are too large, the offset will rise too high above the obstacle. If the bends are made with angles that are too small, the offset will be too low to clear the obstacle.

### Bend Corrections

There are occasions when an error has been made in bending conduit. Some of these errors can be repaired quickly and easily. For example, a bend may be bent past the desired angle or a dogleg may be present in an offset.

If a bend is made past the desired angle, it can be unbent slightly. The conduit is placed back in the bender with the bender shoe in the air. The bender hook is used to hold one leg of the offset, with the bend itself on the curved shoe. The other leg is used as a lever to apply pressure in the opposite direction.

A dogleg can be straightened by placing the bender on the conduit with one of the bends at a right angle to the shoe. Light pressure is applied to bring the bend back in line with the other bend. **See Figure 3-13.**

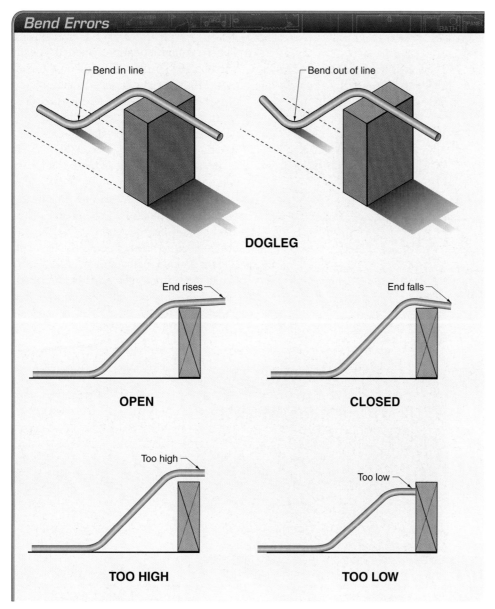

Bend Errors

Bend in line

Bend out of line

**DOGLEG**

End rises

End falls

**OPEN**

**CLOSED**

Too high

Too low

**TOO HIGH**

**TOO LOW**

*Figure 3-12. Common errors in offset bends include a dogleg, open or closed bends, or a bend that is too high or too low.*

## Box Offsets

A *box offset* is a small offset bend used to move conduit away from a wall and into an electrical box. A box offset is typically made with two 5° or 10° bends that rise about ½″. **See Figure 3-14.**

Since most modern benders have a 10° angle marking, this is the most common bend angle. Most often, box offsets are made by eye. However, very accurate and consistent box offsets can be made as follows:

1. Place a pencil mark 2¾″ from the end of the conduit.

2. Turn bender upside down with the handle on the floor. Place the conduit in the bender and align the end of the conduit to the edge of the hook. Make a 10° bend.

3. Rotate the conduit 180° and slide the conduit in the bender to align the pencil mark with the edge of the hook. Make the other 10° bend.

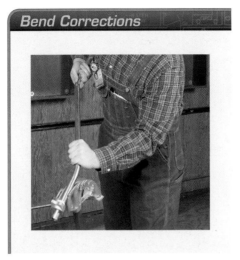

**Bend Corrections**

**Figure 3-13.** *Doglegs are easily corrected by turning the conduit 90° in the bender and applying light pressure.*

## Offset Bend Pre-Positioning

There are many situations where the bends must be at a specific location on the conduit. This is often the case when conduit must be worked around an obstruction or when the bends of several conduits must be aligned to obtain a neat and workmanlike result.

If an obstacle is near the middle of a conduit, the conduit should be bent so that it correctly joins with the previous length of conduit or junction box and so that the last bend of the offset narrowly clears the obstacle. This can be accomplished by cutting the conduit to fit after the offset bend is made. However, this may lead to increased labor costs and waste of conduit.

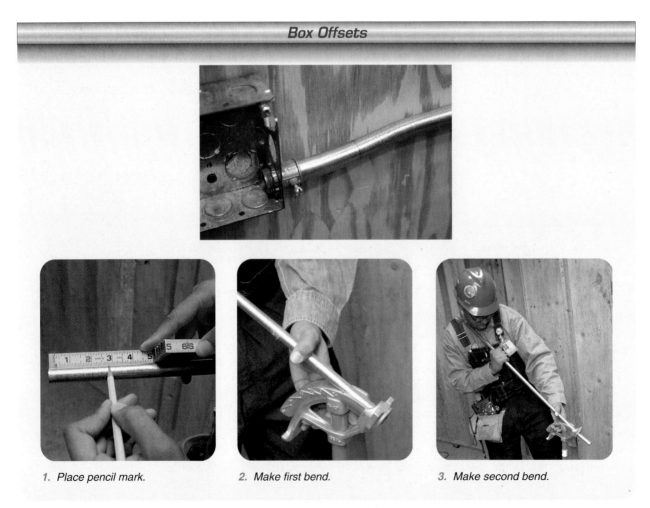

**Box Offsets**

1. Place pencil mark.  2. Make first bend.  3. Make second bend.

**Figure 3-14.** *A box offset is a small offset bend used to move conduit away from a wall and into an electrical box.*

**Calculating Shrink.** There are some simple calculations used to determine where to place the pencil marks on the conduit to ensure that the bends are made in the right place. When conduit is bent around an obstacle, the conduit "shrinks." **See Figure 3-15.**

A *shrink constant* is the reduction in distance that a conduit can run per inch of offset elevation. The conduit itself does not shrink; instead, the total run is reduced. *Shrink* is the amount by which the total run that conduit can cover is reduced because of the extra length required to bend around an obstacle.

In other words, the run that a conduit can cover is reduced from its original length because some of that length is used up in rising above an obstacle. Shrink and the shrink constant are related as follows:

*Shrink = constant × offset rise*

$$S = C \times O$$

where

$S$ = shrink, in inches
$C$ = shrink constant, in inches per inch of offset
$O$ = offset rise, in inches

For hand bending, any bend angle has a specific shrink constant. Most often, the shrink constant is obtained from a table. **See Figure 3-16.** It is useful to note that a very common bend, 30°, has a distance multiplier of 2.0 and a shrink constant of about ¼. These numbers are easy to remember, easy to use, and help to account for the popularity of the 30° bend.

For example, a 30° bend is chosen to go over a 4½″ (4.5″) obstacle. The shrink constant from the table is ¼. The shrink is calculated as follows:

$$S = C \times O$$
$$S = ¼ \times 4.5$$
$$S = \mathbf{1.125″, \ or \ 1⅛″}$$

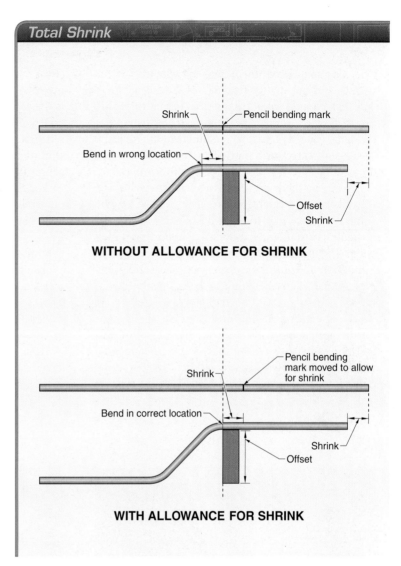

**Figure 3-15.** *The run that a conduit can cover shrinks as the conduit is bent around an obstacle.*

| Calculating Shrink | |
|---|---|
| Bend Angle, θ | Shrink Constant |
| 5° | 0.044 ≈ ¹⁄₁₆ |
| 10° | 0.087 ≈ ¹⁄₁₆ |
| 15° | 0.13 ≈ ⅛ |
| 22½° | 0.20 ≈ ³⁄₁₆ |
| 30° | 0.27 ≈ ¼ |
| 45° | 0.41 ≈ ⅜ |

**Figure 3-16.** *A table of shrink constants for standard angles is used to calculate the shrink.*

## Rolling Offsets . . .

There are some occasions where an offset must move in the horizontal and vertical planes at the same time. This can happen when conduit that is running along a wall has to rise and move away from the wall at the same time. The easiest method is to measure the distance directly and simply rotate the conduit when connecting it to the previous point of termination.

However, there are cases where measuring the offset is not so straightforward. The conduit may be some distance away from its destination, it may be difficult to install and temporarily support, or it may need to be installed in a confined or crowded area. In a situation like this, it may be easier to use additional measurement and calculation to determine the overall offset needed.

The amount of offset is the distance from the conduit to the opening in the enclosure. The actual vertical and horizontal rise can be shown as a triangle. One side of the triangle (A) is the vertical rise from the conduit to the opening. The other side of the triangle (B) is the horizontal rise from the conduit to the opening. The length of the hypotenuse of the triangle (C) is the required offset.

It is easiest to measure from the floor and from the wall. However, the conduit is above the floor and slightly away from the wall. Therefore, the distances can be calculated by taking the differences.

For example, the front of the connector is $4\frac{3}{16}''$ from the wall and the front of the conduit is $2\frac{1}{16}''$ from the wall. Therefore, the difference between these numbers, $2\frac{1}{8}''$, is the horizontal distance from the front of the conduit to the front of the connector.

The top of the connector is $23\frac{5}{16}''$ from the floor and the top of the conduit is $15\frac{7}{16}''$ from the floor. Therefore, the difference between these numbers, $7\frac{7}{8}''$, is the vertical distance from the top of the conduit to the top of the connector.

Both sides of the triangle are now known. The Pythagorean theorem states that, for a right triangle, the sum of the sides squared equals the hypotenuse squared.

The hypotenuse of the triangle is the actual offset rise. This is stated in a formula as follows:

$$A^2 + B^2 = C^2$$

or

$$C = \sqrt{A^2 + B^2}$$

where

$A$ = length of one side, in inches
$B$ = length of other side, in inches
$C$ = length of hypotenuse, in inches

In this case, $A = 7\frac{7}{8}''$ (7.875″) and $B = 2\frac{1}{8}''$ (2.125″). The hypotenuse, or offset, can be calculated as follows:

$$C = \sqrt{A^2 + B^2}$$
$$C = \sqrt{7.875^2 + 2.125^2}$$
$$C = \sqrt{62.016 + 4.516}$$
$$C = \sqrt{66.532}$$
$$C = \mathbf{8.16''} \text{ or } \mathbf{8\frac{3}{16}''}$$

The calculated offset is 8³⁄₁₆″. This matches the original direct measurement of the offset. At times, it may be necessary to establish measuring reference lines in order to determine the measurement in each axis between the conduit and the connector, knockout, or coupling where the conduit will terminate. The electrician may need to use mortar joints, plumb lines, levels, lasers, or other devices to establish these reference lines.

**Determining Precise Offset Locations Using Shrink.** Once the amount of shrink is calculated, the location of the bends can be determined. When working away from a fixed location, such as a junction box or the end of the previous length of conduit, the procedure is as follows:

1. Measure the offset. Choose the desired bend angle. Calculate the shrink. Calculate the distance between bends. **See Figure 3-17.**

2. Measure the distance from a fixed location to the obstacle. This distance is the flat linear distance from the location of the fixed end of the conduit to the beginning of the obstruction. Add the amount of total shrink to the measured distance. This new value is the distance from the fixed location to the end of the first bend.

3. On the conduit, measure out to the value calculated above. Make a pencil mark on the conduit at that point. Mark 1 is the first bend location. Measure the calculated distance between bends back from the pencil mark toward the fixed end. Make a pencil mark on the conduit at that point. Mark 2 is the second bend location.

4. Place the conduit in the bender and align the arrow mark of the bender with the first mark. Bend back toward the fixed end to the desired angle. From this point, the bends are made in exactly the same method as described earlier.

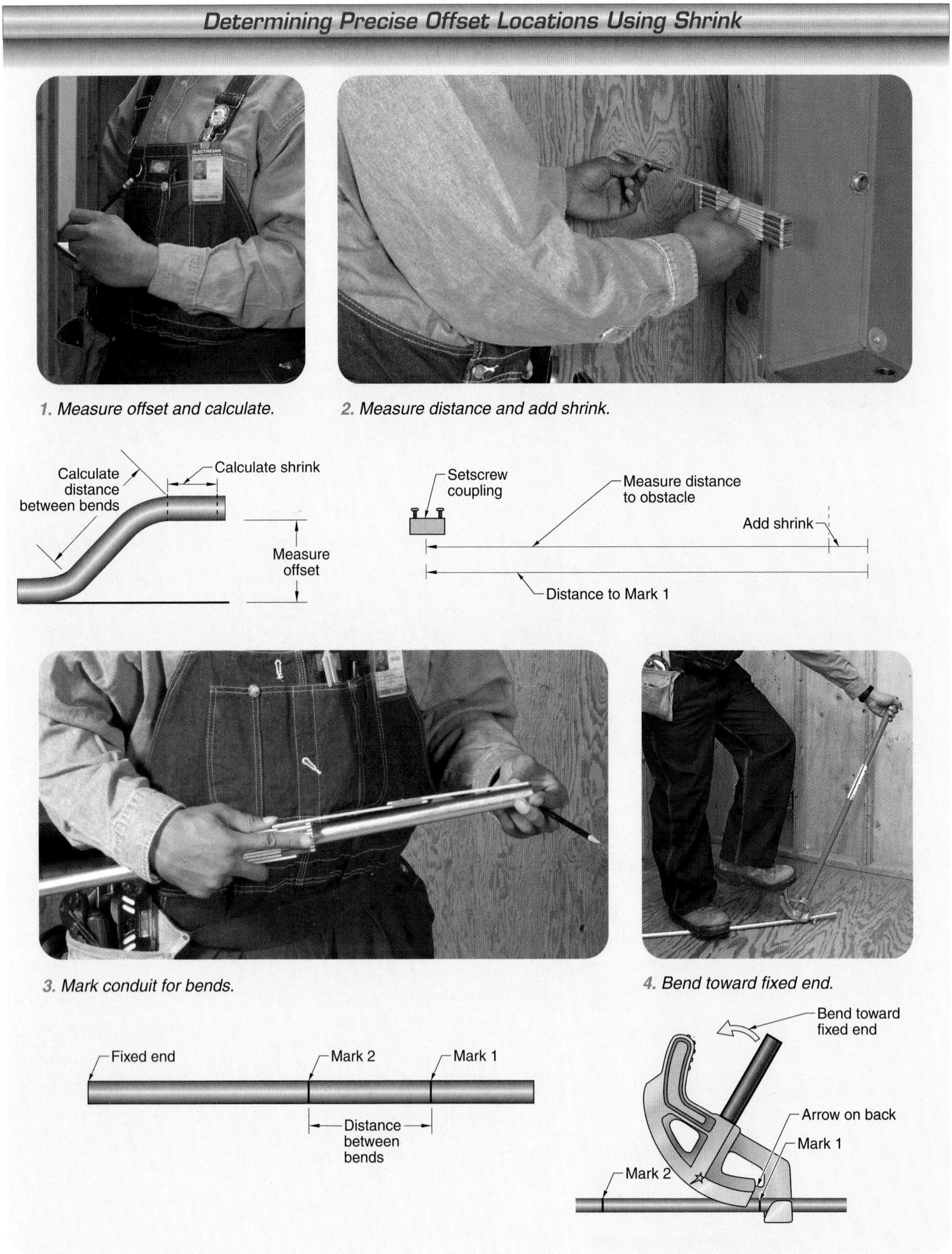

## Determining Precise Offset Locations Using Shrink

1. Measure offset and calculate.

2. Measure distance and add shrink.

Calculate distance between bends

Calculate shrink

Measure offset

Setscrew coupling

Measure distance to obstacle

Add shrink

Distance to Mark 1

3. Mark conduit for bends.

4. Bend toward fixed end.

Fixed end

Mark 2

Mark 1

Distance between bends

Bend toward fixed end

Arrow on back

Mark 1

Mark 2

**Figure 3-17.** Shrink is used to mark the precise location of offset bends.

**Application—Offset around Duct with Shrink.** A run of ¾″ EMT is being run along the ceiling. A rectangular ventilation duct is in the way of the run and an offset must be bent around the duct. The offset rise and distance to the duct need to be measured. **See Figure 3-18.**

In step 1, the duct is measured at 9⅝″ (9.625″). Since there is limited room for the offset, a 45° angle is chosen for the bend. For a 45° angle, the constant is 0.41″ per inch of offset. The shrink is calculated as follows:

$$S = C \times O$$

$$S = 0.41 \times 9.625$$

$$S = \mathbf{3.95″, \ or \ 3^{15}\!/_{16}″}$$

For a 45° bend, the multiplier for distance between bends is 1.41. The distance between bends is calculated as follows:

$$D = M \times O$$

$$D = 1.41 \times 9.625$$

$$D = \mathbf{13.57″, \ or \ 13^{9}\!/_{16}″}$$

In step 2, the distance from the fixed end of the conduit to the duct is measured as 57⅞₁₆″ (57.44″). **See Figure 3-19.** The shrink is added to the measured distance to the obstacle. This new distance is 61⅜″ (57.44 + 3.95 = 61.39, or 61⅜).

In step 3, the first pencil mark is made at a distance of 61⅜″ from the fixed end. The calculated distance between bends is 13⅀₁₆″. This amount is measured back toward the fixed end of the conduit and the second pencil mark is made. This is equivalent to measuring forward 47¹³⁄₁₆″ from the fixed end (61⅜″ − 13⁹⁄₁₆″ = 47¹³⁄₁₆″).

In step 4, the bender is then placed on the conduit and the arrow mark on the bender is aligned with the first pencil mark. The bender is placed on the conduit with the hook facing the duct, away from the fixed end. The conduit is bent back toward the fixed end to the desired angle. This completes the first bend.

The bender is turned upside down and the conduit moved in the bender until the second mark is aligned with the arrow on the bender. The conduit is rotated 180° in the bender and the second bend started. The second bend is finished on the floor.

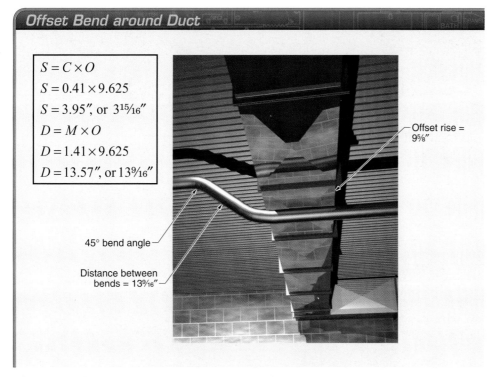

**Offset Bend around Duct**

$$S = C \times O$$
$$S = 0.41 \times 9.625$$
$$S = 3.95″, \ or \ 3^{15}\!/_{16}″$$
$$D = M \times O$$
$$D = 1.41 \times 9.625$$
$$D = 13.57″, \ or \ 13^{9}\!/_{16}″$$

Offset rise = 9⅝″

45° bend angle

Distance between bends = 13⁹⁄₁₆″

*Figure 3-18. When bending around ductwork, the offset and distance to the duct must be measured.*

## Calculating Shrink with Nonstandard Angles

There are situations where a bend must be made at a nonstandard angle. The shrink value can be determined from the triangle created by the offset bend.

The total run of conduit after the bends are made includes the length of the hypotenuse ($H$). The original distance that the conduit could have run includes the length of the adjacent side ($A$). The difference between these two lengths is the total shrink. This is calculated as follows:

$$S = H - A$$

where

$S$ = shrink, in inches
$H$ = length of hypotenuse of triangle, in inches
$A$ = length of adjacent side of triangle, in inches

Since the length of the hypotenuse is not readily known, this equation can be restated using trigonometry as follows:

$$S = (\csc(\theta) - \cot(\theta)) \times O$$
$$S = C \times O$$

where

$S$ = shrink, in inches
$\theta$ = bend angle, in degrees
$\csc(\theta)$ = cosecant of the bend angle
$\cot(\theta)$ = cotangent of the bend angle
$O$ = offset rise, in inches
$C$ = shrink constant

The value within the parentheses, $(\csc(\theta) - \cot(\theta))$, is the shrink value, available in tables for standard angles. However, this must be calculated for nonstandard angles. The values of $\csc(\theta)$ and $\cot(\theta)$ are available in trigonometry tables. **See Appendix.**

For example, a bend must be made with a 35° angle because of multiple bends around obstacles. The offset rise is 9″. The shrink is calculated as follows:

$$S = (\csc(\theta) - \cot(\theta)) \times O$$
$$S = (\csc(35) - \cot(35)) \times 9$$

From the trigonometry table, the value of csc(35) is 1.74 and the value of cot(35) is 1.43. **See Appendix.** The calculation is completed as follows:

$$S = (1.74 - 1.43) \times 9$$
$$S = (0.31) \times 9$$
$$S = \mathbf{2.79″}$$

This calculation can be verified by comparing the value of the shrink determined here with the value from the table for similar angles. For a 35° angle, the angle is between 30° and 45°. The shrink values for these angles are 0.25 of 0.35, respectively. The calculated shrink value of 0.31 is between these values. This indicates that the calculation was done correctly.

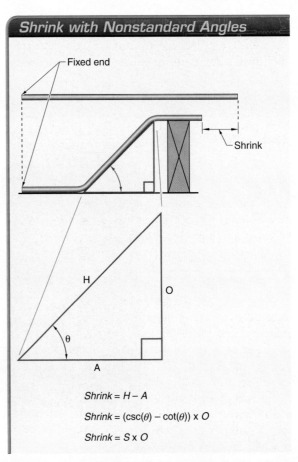

For nonstandard angles, the shrink value can be determined from the triangle created by the offset bend.

**Offset Bend—Measurement to the Obstacle**

$$\begin{array}{r} 57\frac{7}{16}'' \\ + 3\frac{15}{16}'' \\ \hline 61\frac{3}{8}'' \end{array}$$

Distance to edge = $57\frac{7}{16}''$

Measure back $13\frac{9}{16}''$ for second mark

Pencil mark at $61\frac{3}{8}''$ from end

**Figure 3-19.** *The shrink is added to the distance to the obstacle to determine where to place the bend.*

## Parallel Offset Bends

Offset bends can pose a spacing problem when run in a parallel configuration such as in a conduit rack out of a panel or junction box. Conduits that are spaced evenly will crowd one another at the angled sections of the offsets. **See Figure 3-20.**

For conduit that is exposed to view, special techniques are needed to ensure that the conduits remain evenly spaced throughout the bend. The position of the bends in adjacent lengths of conduit must be adjusted to maintain equal spacing between offsets. As the offsets in each adjacent length of conduit are adjusted relative to the original conduit, the spacing between the offsets will remain equal at the angled sections. If this adjustment is not made, the offsets will be too close together and will look out of proportion.

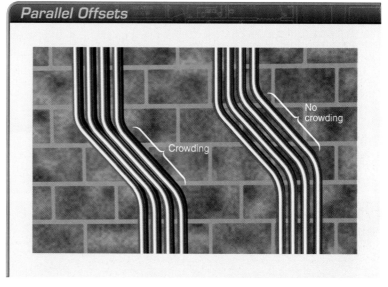

**Parallel Offsets**

No crowding

Crowding

**Figure 3-20.** *Parallel offset bends crowd each other when spacing adjustments are not made.*

The direction of the adjustment depends on the direction of the bends and whether the electrician is working from right to left or left to right. The direction of the adjustment can be determined by inspection of the layout.

For example, when the offset is bent to the left and when working from right to left, the center of the bends must move down relative to the offset that was bent on the first conduit. This means that the pencil marks on the conduit are moved down by the calculated amount of the adjustment relative to the offset that was bent on the first conduit.

When working from left to right with the offset bend to the left, the center of the bends must move up relative to the first conduit. This means that the pencil marks are moved up by the calculated amount of the adjustment.

*Offsets can be used to change the plane of a conduit run.*

The amount of adjustment in the layout of the bends is determined simply by multiplying the distance between centerlines of the adjacent conduits by a constant. The adjustment is calculated as follows:

*adjustment = constant × distance between centerlines*

$$A = C \times D$$

where

$A$ = adjustment, in inches
$C$ = constant
$D$ = distance between conduit centerlines, in inches

The parallel offset constant depends on the bend angle and is available in a parallel offset constant table. **See Figure 3-21.** Note that the parallel offset constant has the same value as the shrink constant.

For example, three conduits need to be offset to the left and in parallel. The conduit is ½″ EMT, placed 2″ on center with a 45° offset bend angle.

The first conduit is bent in the normal way for offset bends. The bends in the second conduit must be adjusted to keep the two conduits evenly spaced at the offset. From the table, the constant for a 45° bend is 0.414. The amount of adjustment is calculated as follows:

$$A = C \times D$$
$$A = 0.414 \times 2$$
$$A = \mathbf{0.828''}, \text{ or } \mathbf{^{13}\!/_{16}''}$$

| Calculating Parallel Offsets | |
|---|---|
| Bend Angle, θ | Parallel Offset Constant |
| 5° | 0.044 ≈ ¹⁄₁₆ |
| 10° | 0.087 ≈ ¹⁄₁₆ |
| 15° | 0.132 ≈ ⅛ |
| 22½° | 0.199 ≈ ³⁄₁₆ |
| 30° | 0.268 ≈ ¼ |
| 45° | 0.414 ≈ ⅜ |

**Figure 3-21.** *The shift value for parallel offsets is used to adjust the location of adjacent conduit with parallel offset bends.*

When working from right to left, the pencil marks on the second conduit must be adjusted by moving them down ¹³⁄₁₆″ relative to the first conduit. **See Figure 3-22.** The pencil marks in the third conduit are adjusted by moving them down by ¹³⁄₁₆″ relative to the second conduit or 1⅝″ from the first conduit. When the bends are complete, the centers of the bends align on a line at an angle equal to half the bend angle. The offsets are equally spaced, look uniform, and are not crowded.

**Application—Parallel Offsets on a Conduit Rack.** Three conduits must be run out of an enclosure in parallel and bent in parallel offsets. **See Figure 3-23.** The conduit is 1″ EMT, placed 6″ on center, with a 30° bend angle.

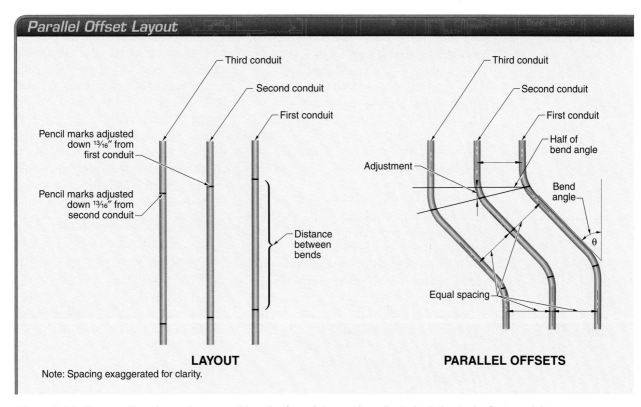

**Parallel Offset Layout**

Third conduit
Second conduit
First conduit

Pencil marks adjusted down ¹³⁄₁₆″ from first conduit

Pencil marks adjusted down ¹³⁄₁₆″ from second conduit

Distance between bends

**LAYOUT**

Note: Spacing exaggerated for clarity.

Third conduit
Second conduit
First conduit

Half of bend angle

Adjustment

Bend angle

θ

Equal spacing

**PARALLEL OFFSETS**

*Figure 3-22. The pencil marks on the second length of conduit must be adjusted relative to the first conduit.*

The first conduit is measured and bent in the normal way for offsets. The location of the bends in the second conduit must be adjusted to keep the two conduits evenly spaced through the offsets. From the parallel offset constant table, the constant for a 30° bend is 0.268. The amount of adjustment is calculated as follows:

$$A = C \times D$$
$$A = 0.268 \times 6$$
$$A = \textbf{1.61}'', \text{ or } \textbf{1}⅝''$$

The pencil marks on the second conduit must be shifted 1⅝″ relative to the first conduit. Since the bends are to the right, and the electrician is working from right to left, the pencil marks are shifted away from the enclosure so the bends are farther from the enclosure. For the third conduit, the pencil marks must be shifted another 1⅝″ in the same direction relative to the second conduit. This is the same as shifting the marks 3¼″ relative to the first conduit.

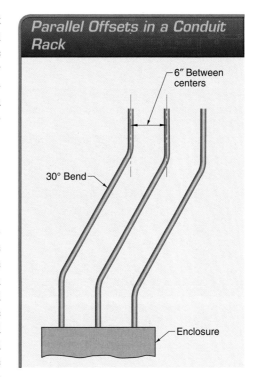

**Parallel Offsets in a Conduit Rack**

6″ Between centers

30° Bend

Enclosure

*Figure 3-23. The location of the bends on adjacent lengths of conduit must be adjusted to modify the distance between the centerlines.*

## Measured Rise Method for Offset Bends

The measured rise method produces offsets using a method that does not require the electrician to fabricate conduit bends to a specific angle. While this method is rarely used with hand benders, where known angles can generally be bent accurately, this technique has value when using mechanical and hydraulic benders.

**Making the Bends.** The techniques used to lay out and fabricate the offset are unique to the measured rise method. The offset is fabricated as follows:

1. Mark the location of the first bend on the conduit with a pencil mark. Place the bender on the conduit aligning the pencil mark with an easily remembered bender benchmark. Bend the conduit to an appropriate angle. **See Figure 3-24.** Remove the conduit from the bender and place it on a flat surface.

2. Find the center of the bend by extending (with a pencil mark) both of the inside edges of the conduit through the bend. The point at which the two marks intersect is the center of the unknown angle.

3. Measure the distance along the centerline of the conduit from the center of the bend to the mark placed on the conduit (the mark that was aligned with the bender benchmark). This measurement is the take-up for this particular bend on this particular bender.

After the first bend is fabricated, the center of the bend found, and the take-up measured, the second bend is laid out on the floor. **See Figure 3-25.** The second bend is laid out and fabricated as follows:

1. Place the back leading section of the conduit against a straightedge such as the bender handle. Use a rule or framing square held against the straightedge.

2. Slide the rule until the desired amount of rise (on the rule) intersects the conduit edge closest to the straightedge. Place a mark on the conduit at this location. This is the center of the second bend.

3. From this mark, add, in the same direction, the take-up value measured at the first bend. This will be where the bender benchmark will be placed for the second bend.

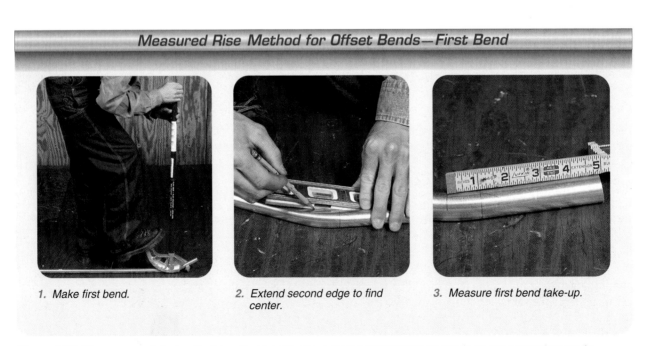

**Measured Rise Method for Offset Bends—First Bend**

1. Make first bend.

2. Extend second edge to find center.

3. Measure first bend take-up.

**Figure 3-24.** *The measured rise method for offset bends allows for the first bend to be made at any convenient angle.*

4. Place the bender on the conduit and align the same bender benchmark with the proper conduit mark. Turn the bender upside-down, align the handle on the same plane with the first bend, and start the second bend in the air.

5. Once the end of the first bend disappears from view when sighting down the conduit, turn the bender around and complete the bend on the floor.

6. Measure the offset rise to verify that the bend was made correctly.

The bend is completed when the leading section of the conduit is parallel to the trailing section of the offset. If the working surface is level, the two sections of the offset will both be level and parallel to the floor.

Using the measured rise method for making bends can be more time consuming than using the multiplier method. Since the bend angles are not standardized, it can be very difficult to make parallel bends. In addition, it also makes it very difficult for another electrician to match the bends if adding conduit to the run at a later date.

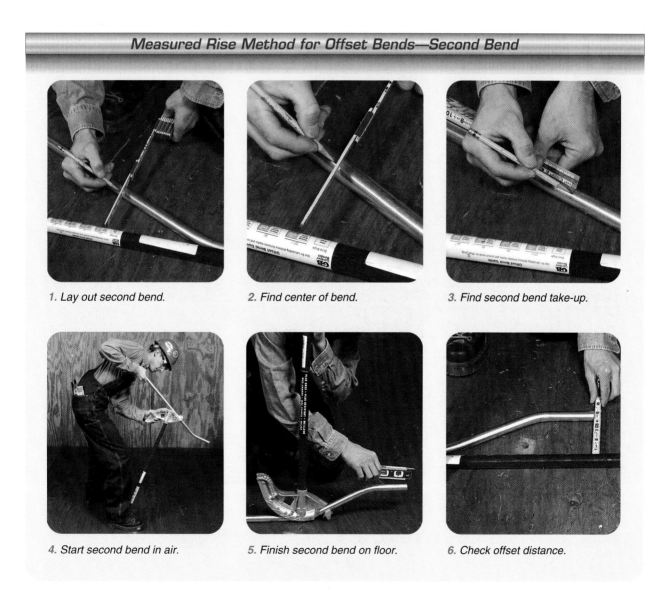

**Measured Rise Method for Offset Bends—Second Bend**

*1. Lay out second bend.*

*2. Find center of bend.*

*3. Find second bend take-up.*

*4. Start second bend in air.*

*5. Finish second bend on floor.*

*6. Check offset distance.*

**Figure 3-25.** *The second bend in the measured rise method is measured from the center of the first bend.*

## KICKS

A *kick* is any bend less than 90° that is used to change direction in a conduit run. **See Figure 3-26.** Kicks are generally used to move a 90° bend away from a wall, toward a wall, or to go around an obstruction. Kicks fulfill a similar function as offsets. In most circumstances, kicks are easier to make than offsets because they require only one angled bend.

Kicks

**Figure 3-26.** *A kick is any bend less than 90° that is used to change direction in a conduit run.*

The three variables to consider when making kicks are the amount of rise needed to clear the obstruction, the optimum location on the conduit run to make the kick, and the bend angle of the kick. The bend angle is generally only considered when running a rack of several conduits.

Kicks are closely related to offsets and can be made using two different methods. The measured rise method only measures the amount of rise as the bend is made. The multiplier method uses the cosecant of the known angle as the multiplier. This method is often used in conduit rack installations where it is highly desirable for all the kicks on that rack have the same degree of bend.

### Measured Rise Method for Kicks

A kick made with the measured rise method is one of the easiest of all bends to fabricate. **See Figure 3-27.** The procedure to make kicks is as follows:

1. Measure the amount of rise necessary (top to top or bottom to bottom).

2. Place the bender at the proper location on the conduit, making certain the stub of the 90° bend is at 90° to the bender handle and turned in the correct direction. Slowly begin bending the conduit using heavy foot pressure. At the same time, hold a ruler tight to the floor and measure the height to the bottom of the 90° stub.

3. Once the desired amount of rise is reached, stop bending and remove the bender from the conduit. Check the measurement by holding the 90° stub level and measuring from the floor to the bottom of the conduit. Some minor adjustment may be necessary to fine-tune the kick.

### Multiplier Method for Kicks

This method is very similar to the multiplier method used for offset bends. This method treats the kick as an offset, but the nature of kicks makes the calculation a little more difficult. The multiplier method is often used when running several conduits on a rack. In this situation, keeping the angles consistent from conduit to conduit is desirable. The multiplier method is used as follows:

1. Measure the required rise of the kick. Fabricate the 90° bend to the proper length. Choose the angle for the kick that best suits the situation. Use a straight edge to extend the back of the 90° bend. **See Figure 3-28.**

2. Calculate the product of the required rise and the cosecant of the chosen kick angle. This is the distance from the back of the developed 90° bend to the center of the bend used to produce the kick. Measure this distance from the back of the 90° bend and place a pencil mark on the conduit.

3. If a 45° bend angle is chosen, align the pencil mark with the star benchmark for the center of that bend.

4. Use heavy foot pressure to make the bend. If necessary, measure the angle with a level or protractor.

The 90° bend must be turned in the correct direction, and the desired angle for the bend should be made using heavy foot pressure. The rise of the kick should be the proper length and every other kick on the rack can be bent with the same angle using the same method.

If the desired angle is not 45° or the available bender does not have the proper marks, a piece of scrap conduit must be used to find the center of the bend for this particular angle. This is accomplished as follows:

1. On a straight piece of scrap conduit, make a pencil mark, align the bender's arrow benchmark with the pencil mark, and bend the conduit to the desired angle.

2. Remove the scrap conduit from the bender and locate the center of the bend by extending the inside of the leading and trailing sections of the bend into the bend with a pencil mark.

3. Place the conduit back in the bender shoe and transfer the center mark to the shoe. This will be the center of the angle. Place a pencil mark where the lines meet as a new benchmark. This is the center of the bend.

4. Place the bender on the conduit in which the kick is to be formed and align the new benchmark with the pencil mark on the conduit.

5. Bend the conduit to the desired angle. The degree of bend can be checked with a level and the overall rise of the kick can be verified after the bend is complete.

## Measured Rise Method for Kicks

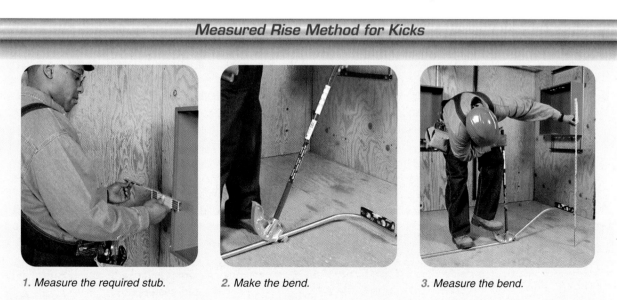

*1. Measure the required stub.*     *2. Make the bend.*     *3. Measure the bend.*

**Figure 3-27.** *The measured rise method for kicks is a simple method for bending around an obstacle.*

**Multiplier Method for Kicks**

*1. Fabricate the 90° bend.*

*2. Mark the conduit.*

*3. Align pencil mark with star or center-of-bend benchmark.*

*4. Bend kick.*

**Figure 3-28.** *The multiplier method for kicks is very similar to the multiplier method for offset bends.*

If the rack consists of many different sizes of conduit, the center of bend for that angle should be found for every bender required to complete the rack. The multiplier method is especially useful when extremely accurate electric benders are available. With this procedure the rise can be calculated for any degree of bend.

## APPLICATION— PARALLEL OFFSETS WITH CONDUIT OF DIFFERENT SIZES

The best parallel offset bends are made with conduit of the same size and using the same

bender throughout. However, acceptable results can be obtained when bending parallel offsets in conduit of different sizes and using different benders, provided the conduits do not vary greatly in size. **See Figure 3-29.**

As long as each conduit is within a size or two of the adjacent conduits, the bends can be laid out in a similar method to what would be done if all the conduits were the same size. Care must be taken to accurately determine the distance between centerlines and to adust the position of the offset in each conduit.

The key to efficiently bending parallel offsets is to make the parallel offset bends at the center of the bends instead of at the arrow benchmark. The center of the bends can be found by making a bend on a scrap piece of conduit and permanently transferring the mark to the bender shoe as a new benchmark.

Establishing bender benchmarks for the center of standard bends, such as 15°, 22½°, and 30°, ultimately saves time because the marks are permanent and never have to be located again. Once the bender is set up with benchmarks for the desired standard angles, parallel offsets can be made for conduit of different sizes.

To make the parallel offsets, one of the conduits is used as a reference, or baseline. Once the center of bend location is known for the first conduit, the bend center locations on the subsequent conduits are adjusted using the parallel offset calculations.

**Application—Parallel Offsets**

**Figure 3-29.** *Parallel offsets with conduit of different sizes presents unique challenges.*

 **Calculating Parallel Offset Adjustments**

A close inspection of parallel offsets shows that the locations of the bend centers are adjusted by an angle equal to half the bend angle. Applying trigonometry, the constant used to calculate the adjustment is the tangent of that half angle. The constant is calculated as follows:

*constant = tan (half of bend angle)*

$$C = \tan\left(\frac{\theta}{2}\right)$$

where

$C$ = constant
$\theta$ = bend angle

For example, for a 45° bend angle, half the angle is 22½°. The tangent of 22½° is 0.414. Therefore, the constant for a 45° bend angle is 0.414. This matches the value in the table and this shows that the constant can be calculated in the field with a scientific calculator. When the bend angle and the distance between centers is known, the adjustment is calculated as follows:

*adjustment = distance between centerlines ×
tan (half of bend angle)*

where

$A$ = adjustment, in inches
$\theta$ = bend angle
$D$ = distance between centerlines, in inches

For example, 3 conduits of ½″ EMT, placed 3¾″ (3.75″) on center, with a 30° bend angle, need to be bent in a parallel offset. The first conduit is measured and bent in the normal way for offsets. The location of the bends in the second conduit must be adjusted to maintain even spacing between the conduits. The amount of the adjustment is calculated as follows:

$$A = \tan\left(\frac{\theta}{2}\right) \times D$$

$$A = \tan\left(\frac{30}{2}\right) \times 3.75$$

$$A = \tan(15) \times 3.75$$

$$A = 0.268 \times 3.75$$

$$A = \mathbf{1.005″}, \text{ or } \mathbf{1″}$$

The pencil marks on the second conduit are moved 1″ relative to the first conduit. The pencil marks on the third conduit are moved another 1″ relative to the second conduit.

## SUMMARY

- An offset bend is a double conduit bend formed with two equal angles in opposite directions that is used to make a bend in a conduit run.

- The multiplier method of fabricating offsets is commonly used to bend offsets at a known angle.

- The distance between offset bends is calculated from a table of multipliers. **See Figure 3-30.**

- The location of the two offset bends is calculated based on the distance to the obstacle and the angle chosen for the bend.

- Shrink is the amount the run length is reduced when conduit is bent around an obstacle.

- Shrink is the product of the shrink constant and the offset rise.

- For parallel offsets that are exposed to view, the bends must be adjusted relative to the first conduit.

- The bend adjustment for parallel offsets is the product of the parallel offset constant and the distance between centerlines of the adjacent conduits.

- The measured rise method of making offset bends is a method of making offset bends that does not require bending the conduit to a specific angle.

- A kick is a bend used to change direction in a conduit run.

- Parallel offsets can be made with conduit of different sizes by marking and bending at the center of each bend.

| Distance Multiplier and Shrink Constant | | |
|---|---|---|
| Bend Angle, θ | Distance Multiplier | Shrink Constant |
| 5° | 11.4 | 0.044 |
| 10° | 5.76 | 0.087 |
| 15° | 3.86 | 0.13 |
| 22½° | 2.61 | 0.20 |
| 30° | 2.00 | 0.27 |
| 45° | 1.41 | 0.41 |

**Figure 3-30.** A table of multipliers for common angles can be used when making the calculations required for offset bends.

# Hand Bending—Saddles and Corner Offsets

## CONDUIT BENDING and FABRICATION

**4**

SADDLES .................................................................................................... 58

THREE-BEND SADDLES .......................................................................... 61

FOUR-BEND SADDLES ........................................................................... 66

CORNER OFFSETS .................................................................................. 69

COMPOUND 90° BENDS .......................................................................... 70

APPLICATION—THREE-BEND SADDLE AROUND A DRAIN ................. 73

SUMMARY ................................................................................................ 76

O ccasionally, a conduit run needs to change elevations to clear an obstruction and then return to its original elevation. The bend required for this is known as a saddle and can be fabricated as either a three-bend or a four-bend saddle. The size and shape of the obstruction determines the type of saddle needed.

There are four variables when making saddle bends: the rise of the obstruction, the width of the obstruction, the distance to the obstruction, and the choice of the bend angles. After the distances are measured and the angle chosen, the shrink and the distance between the bends can be calculated. The shrink and distance between bends are used to determine the placement of the bend marks.

A corner offset is a special type of four-bend fabrication consisting of two offsets turned at 90° relative to each other. Corner offsets are used when a conduit run needs to be placed on the adjoining wall of an inside or an outside corner. Compound 90° bends are used when an object is obstructing a corner and two bends are needed to turn the corner and get past the obstruction.

## OBJECTIVES

1. Describe how to choose which type of saddle is best for a particular situation.
2. List the four variables used when making saddle bends and explain how to determine these variables in the field.
3. Demonstrate how to calculate shrink.
4. Demonstrate how to lay out the bend marks on conduit for three-bend and four-bend saddles.
5. Demonstrate how to fabricate a corner offset.
6. Demonstrate how to fabricate a compound 90° bend.

57

## SADDLES

A *saddle* is a section of conduit consisting of three or four bends that is shaped to bend around an obstacle and then return to its original level. A *three-bend saddle* is a saddle consisting of a center bend and two side bends with the center bend having twice the angle of the side bends. It is essentially two offsets that share a common center bend. **See Figure 4-1.** A three-bend saddle is also called a crossover.

A *four-bend saddle* is a saddle made by placing four bends in a conduit to allow it to move around an obstacle and then return to its original level. A four-bend saddle consists of two offset bends with a length of straight conduit between the offsets.

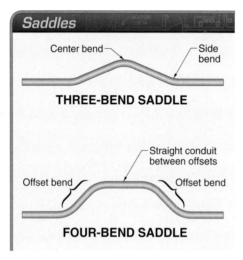

**Figure 4-1.** A three-bend saddle consists of a center bend with two side bends. A four-bend saddle consists of two offsets with a straight length between them.

## Choice of Saddle Type

The choice for the size and type of a saddle bend depends on the size and shape of the obstruction. **See Figure 4-2.** Three-bend saddles are usually used for relatively small cylindrical obstructions where the obstruction is smaller than the bend radius of the center bend. A three-bend saddle can also be used for small square obstructions, typically no more than about 3″ to 4″. However, the fit over a square obstruction will not be as close as the fit over a round obstruction.

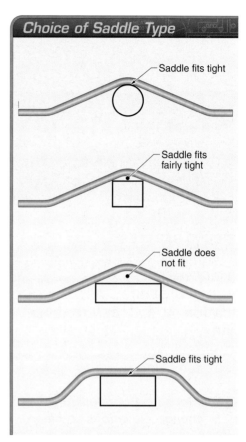

**Figure 4-2.** The choice for the size and type of a saddle bend depends on the size and shape of the obstruction.

The bend radius depends on the conduit size and the bender used. For example, benders for ½″ EMT typically have a radius of just over 4″. This means that when ½″ EMT is bent in a three-bend saddle, it will fit tight around pipe with a diameter of up to 8″. If the pipe is larger, the conduit will not fit correctly. Benders for 1″ EMT have a bend radius of just over 6″. Therefore, 1″ EMT will fit tight around pipe with a diameter of up to 12″.

Three-bend saddles for larger obstructions can be fabricated fairly easily, but they are usually out of proportion to the obstruction and will not fit tightly around it. This can detract from the appearance of the run.

For larger obstructions, a four-bend saddle should be used. A four-bend saddle has a straight section of conduit between the offsets. This straight section can be made any length to cross over rectangular obstructions.

## Saddle Bend Variables

There are four variables used when making saddle bends. These variables are the rise of the obstruction, the width of the obstruction, the distance to the obstruction, and the choice of bend angles. After these variables are determined, the shrink can be calculated.

**Obstruction Rise.** The rise of the obstruction is measured using the same method that is used to measure offset rise. **See Figure 4-3.** The rise is the distance from the bottom of the conduit to the top of the obstruction. If the conduit is mounted directly to a wall or other supporting surface, the obstruction rise can be measured directly from the supporting surface to the top of the obstruction. If the conduit is held by a standoff, the obstruction rise is measured from the bottom or back of the conduit to the top of the obstruction.

For a cylindrical obstruction, such as a sewer pipe or drainage pipe, the top of the obstruction is the center point where the saddle has to cross over the obstruction. For a square or rectangular obstruction, such as an air duct, the top of the obstruction can be measured at the edge where the conduit has to reach the top.

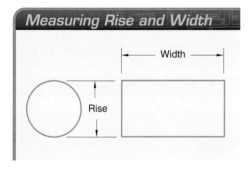

**Figure 4-3.** *The rise is the distance from the bottom of the conduit to the top of the obstruction.*

**Obstruction Width.** The width of the obstruction is the distance from one edge to the other edge. For square and rectangular obstructions, the width is the distance across the object that the conduit has to run. This is the distance between the offsets of a four-bend saddle.

**Distance to Obstruction.** The distance to the obstruction is the distance from the end of the existing conduit run to the obstruction. **See Figure 4-4.** The end of the existing conduit run is the point where the new conduit begins. This may be a coupling, junction box, enclosure, or any other point.

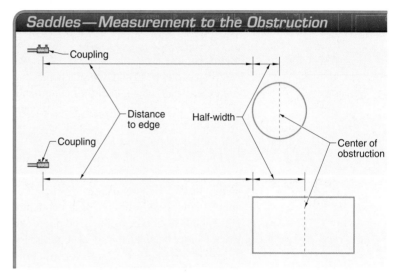

**Figure 4-4.** *The distance to the obstruction is the distance from the end of the existing conduit run to the obstruction.*

For three-bend saddles, the distance is measured to the center of the obstruction. If it is not convenient to measure to the center of the obstruction, the distance to the edge and the width of the obstruction can be measured. The distance to the center is half the width plus the distance to the edge of the obstruction.

For four-bend saddles, there are two common methods of measuring the distance to the obstruction. In the first method, the distance to the edge of the obstruction and the width of the obstruction are used. All bends are made relative to the location of the edge of the obstruction.

---

## Tech Fact

Hydronic piping is almost always covered with insulation. This insulation may not be installed while conduit systems are being run. The thickness of this insulation must be taken into consideration as part of the obstruction rise when calculating bends.

In the second method of making four-bend saddles, the distance to the center of the obstruction and the width of the obstruction are used. All bends are made relative to the location of the center of the obstruction. If it is not convenient to measure to the center of the obstruction, the distance to the edge and the width of the obstruction can be measured. The distance to the center is half the width plus the distance to the edge of the obstruction.

**Bend Angles.** Every three-bend saddle consists of a center bend and two side bends that are exactly half the angle of the center bend. The most common choice is a 45° center bend. Other less-common choices are 30° and 60° for the center bend. **See Figure 4-5.**

For the 45° center bend, the two side bends are 22½° each. For the 30° center bend, the two side bends are 15° each. For a 60° center bend, the two side bends are 30° each. The choice of the bend angle is a compromise between the space required for the bend, the ability to clear the obstruction, and the ease of pulling wire.

Every four-bend saddle consists of two offsets with a straight section of conduit between them. The considerations for choosing the bend angles are the same as for any other offset bend. A smaller bend, such as a 22½° or 30° bend, takes up more space, but makes it easier to pull wire. A larger angle, such as a 45° bend, takes up less space, but makes it more difficult to pull wire.

## Calculating Shrink

The shrink calculation for a saddle is similar to the shrink calculation for an offset bend. The shrink constant used depends on the angle of the bends. For a three-bend saddle with a 45° center bend, the side bends are 22½° and the shrink constant is ³⁄₁₆″ per inch of rise. Shrink is the product of the shrink constant and the offset rise and is calculated as follows:

$$shrink = shrink\ constant \times offset\ rise$$

$$S = C \times O$$

where

$S$ = shrink, in inches
$C$ = shrink constant, in inches per inch of rise
$O$ = offset rise, in inches

Note that this is the amount of shrink when the conduit rises up to the top of an obstruction. There is an equal amount of shrink when the conduit drops back down to its original level. Therefore, the total shrink of a length of conduit is twice the amount calculated here. However, the amount of shrink from the rise is usually the only concern unless it is necessary to know the exact location of the other end of the conduit after the bend is complete.

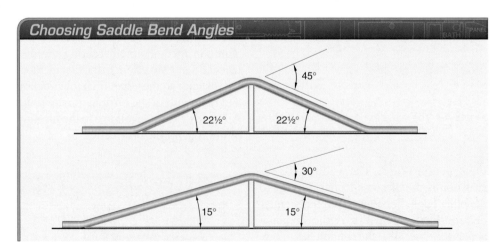

*Choosing Saddle Bend Angles*

**Figure 4-5.** *The most common bend angles are a 45° center bend with two 22½° side bends. A 30° center bend may also be used.*

**Shrink for a Cylindrical Obstruction.** For example, a three-bend saddle is needed to bend conduit around a pipe. **See Figure 4-6.** The rise of the pipe is 4″ and a 45° center bend angle is chosen. For a 45° center bend, the two side bends are 22½°. The constant is ³⁄₁₆″ per inch of rise for a 22½° bend. The shrink is calculated as follows:

$$S = C \times O$$

$$S = \tfrac{3}{16} \times 4$$

$$S = \tfrac{12}{16}$$

$$S = \tfrac{3}{4}″$$

**Shrink for a Rectangular Obstruction.** For a four-bend saddle with a 30° offset bend, the shrink constant is ¼″ per inch of rise. For a 45° offset bend, the shrink constant is ³⁄₈″ per inch of rise.

For example, a four-bend saddle is used to bend conduit around a rectangular obstruction. **See Figure 4-7.** The rise of the obstruction is measured at 6″ and a 30° offset bend angle is chosen. For a 30° offset

bend, the constant is ¼″ per inch of rise. The shrink is calculated as follows:

$$S = C \times O$$

$$S = \tfrac{1}{4} \times 6$$

$$S = \tfrac{6}{4}$$

$$S = \mathbf{1\tfrac{1}{2}″}$$

## THREE-BEND SADDLES

A three-bend saddle is conveniently shaped to fit tightly over round or cylindrical obstructions. A three-bend saddle gives a neat and workmanlike appearance to a run of conduit. However, three-bend saddles take some practice to master.

### Marking the Bend Locations

The four variables of a saddle must be examined before laying out the bends on the conduit. After the measurements are taken and the bend angles chosen, the shrink can be calculated. The next step is to place the pencil marks on the conduit.

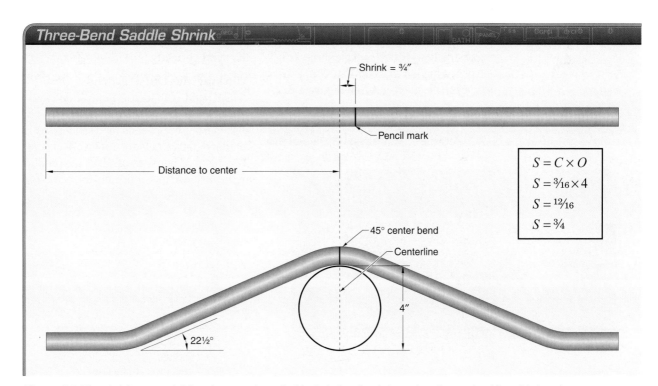

**Three-Bend Saddle Shrink**

Shrink = ¾″

Pencil mark

Distance to center

$$S = C \times O$$
$$S = \tfrac{3}{16} \times 4$$
$$S = \tfrac{12}{16}$$
$$S = \tfrac{3}{4}$$

45° center bend

Centerline

4″

22½°

**Figure 4-6.** *The shrink as conduit bends around a cylindrical obstruction is based on the angle of the side bends.*

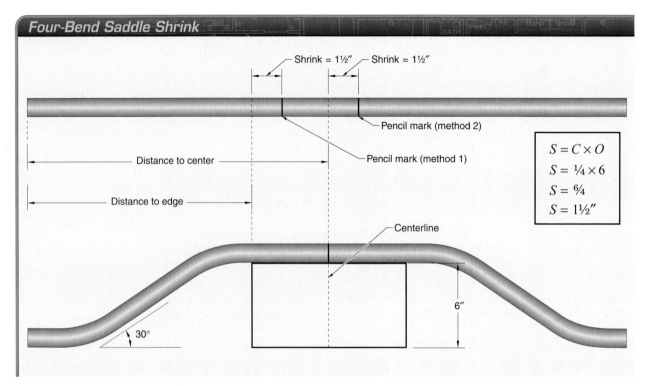

**Figure 4-7.** *The shrink as conduit bends around a rectangular obstruction is based on the angle of the offset bends.*

**Marking the Center Bend.** The location of the center bend pencil mark is determined by adding the shrink to the measured distance to the center of the obstruction. For example, if the distance from the fixed end to the center of the obstruction is 60¼″ and the shrink is ¾″, the pencil mark is placed 61″ (60¼″ + ¾″ = 61″) from the fixed end. **See Figure 4-8.**

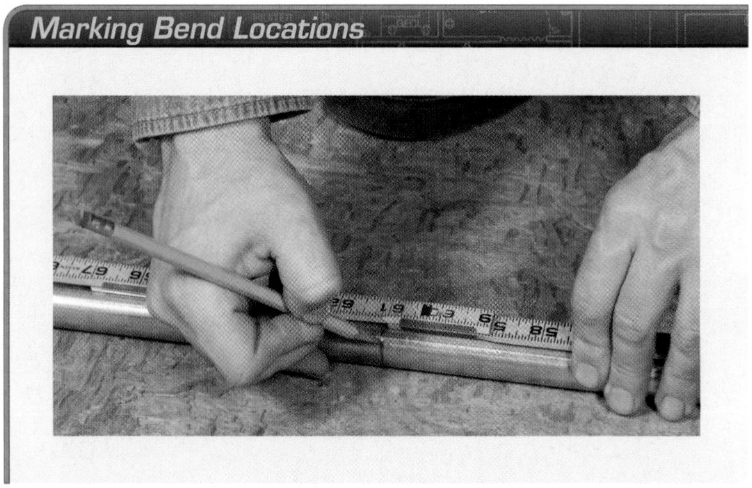

After the saddle is complete, the pencil mark will line up with the center of the obstruction. All other markings are measured from the center pencil mark. The correct placement of this mark is critical to making the bend correctly.

**Marking the Outside Bends.** Once the center bend is marked, the outside angles can be laid out. The calculation for the distance between bends for a saddle is very similar to the calculation for an offset bend. The distance between bends is the product of the distance multiplier and the offset rise and is calculated as follows:

$$distance = distance\ multiplier \times offset\ rise$$

$$D = M \times O$$

where

$D$ = distance between bends, in inches

$O$ = offset rise, in inches

$M$ = distance multiplier

### Marking Bend Locations

**Figure 4-8.** *The location of the center bend pencil mark is determined by adding the shrink to the measured distance to the obstruction.*

The value of the distance multiplier depends on the angle of the bends. **See Figure 4-9.** For a three-bend saddle with a 30° center bend (15° side bends), the distance multiplier is 3.8. For a three-bend saddle with a 45° center bend (22½° side bends), the distance multiplier is 2.5. The side bend pencil marks are placed equal distances on either side of the center bend pencil mark.

| Three-Bend Saddle Multiplier | |
|---|---|
| Side Bend Angle | Distance Multiplier |
| 15° | 3.8 |
| 22½° | 2.5 |

**Figure 4-9.** *The value of the distance multiplier depends on the angle of the bends.*

For example, a three-bend saddle is planned for a pipe with an offset rise of 3½″. A 45° angle is chosen. The distance multiplier is 2.5 and the distance between bends is calculated as follows:

$$D = M \times O$$
$$D = 2\frac{1}{2} \times 3\frac{1}{2}$$
$$D = 8.75'' \text{ or } 8\frac{3}{4}''$$

The pencil marks for the side bends of a three-bend saddle are placed 8¾″ on either side of the pencil mark for the center bend. **See Figure 4-10.** After the bend marks have been laid out on the conduit, the bends can be made.

## Making the Bends

The traditional way of making a three-bend saddle uses the benchmark for the center of a 45° bend for the center bend. The side bends are made with the arrow benchmark with the bender shoe pointing toward the center bend.

Nearly every hand bender has a benchmark representing the center of a 45° bend. This benchmark may be a teardrop or rim notch. If a different bend angle is chosen, a new benchmark must be placed on the bender and used for the bend.

**Making the Center Bend.** The benchmark for the center of the bend is used to make the center bend. A 45° bend is made by placing the bender on the conduit, aligning the center bend pencil mark with the benchmark, and making the bend. **See Figure 4-11.**

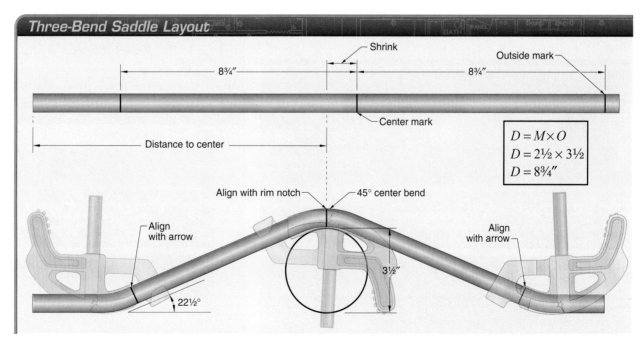

*Three-Bend Saddle Layout*

Shrink

Outside mark

8¾″

8¾″

Center mark

Distance to center

$$D = M \times O$$
$$D = 2\frac{1}{2} \times 3\frac{1}{2}$$
$$D = 8\frac{3}{4}''$$

Align with rim notch

45° center bend

Align with arrow

22½°

3½″

Align with arrow

**Figure 4-10.** *The distance between bends is the product of the rise and the distance multiplier. The pencil marks for the side bends are placed equal distances from the center bend.*

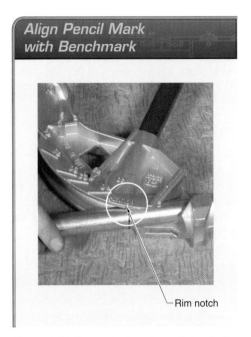

Align Pencil Mark with Benchmark

Rim notch

*Figure 4-11. The center bend is made by placing the bender on the conduit, aligning the pencil mark with the correct benchmark, and making the bend.*

**Making the Second and Third Bends.** The two side bends are made in the air. The bender is placed on the conduit with the bender hook facing the center bend and the pencil mark aligned with the arrow benchmark. **See Figure 4-12.** The conduit is rotated 180° and the first side bend is fabricated.

Bender Hook Direction

Hook points toward center bend

*Figure 4-12. The two outside bends are made in the air with the hook facing the center bend.*

For the second side bend, the bender is removed from the conduit and reversed, and the arrow is aligned with the pencil marks for the other side bend. In other words, the bender hook is pointed at the center bend for both side bends. For a 45° center bend, the two side bends are bent to 22½°. Each new bend must be bent in the same plane as the center bend and the bender handle to avoid creating a dogleg. **See Figure 4-13.**

---

***Tech Fact***

Kinks in a bend reduce the inside diameter and weaken the conduit.

---

Align Bends

*Figure 4-13. Care must be taken to align the bends in the same plane to avoid creating a dogleg.*

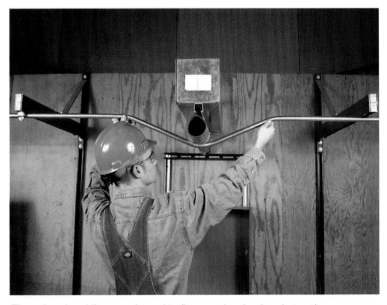

*Three-bend saddles are shaped to fit around a circular obstruction.*

## Push-Through Method

The traditional way of making three-bend saddles requires a different benchmark for the center bend than for the side bends. The rim notch or teardrop benchmark is used for the center bend, and the arrow benchmark is used for the side bends. Because of this, the multiplier for the distance between the bends must be adjusted from what would be expected from the trigonometry of the bends.

For example, a 45° center bend has side bends of 22½°. From the trigonometry, the distance multiplier should be 2.6. However, the value used for the multiplier for three-bend saddles is 2.5. This is to compensate for changing the benchmark from bend to bend.

The push-through method does not require this adjustment to the multiplier. The same multiplier used to fabricate offset bends is used. For example, for a 45° center bend, a multiplier of 2.61 should be used because the side bends are 22½°.

| Offset Table | |
|---|---|
| Side Bend Angle | Distance Multiplier |
| 15° | 3.86 |
| 22½° | 2.61 |

The push-through method requires that new benchmarks be placed for the center of the bend for 30°, 22½°, and 15° bends. These new benchmarks represent the center of the bend for those angles. The process of placing new benchmarks on the bender takes some time, but the long-term benefit is that saddles can be fabricated quickly and accurately.

For the push-through method, the measurements of the size of the obstruction, the distance to the obstruction, and the shrink calculation are made in the normal manner. The distance between bends is calculated with the number in the offset table. In addition, the outside diameter of the conduit is added to the calculated distance. The pencil marks are also made in the normal manner.

In the push-through method, the bends are made starting with one of the outside bends. The bender is placed on the conduit with the benchmark for the center of that bend angle aligned with the pencil mark. All bends are made with the bender pointing toward the same end of the conduit.

For example, a conduit must cross over a 4″ pipe. A 45° center bend angle is chosen. The shrink is calculated for the 22½° side angle. The shrink constant is ³⁄₁₆″ per inch of rise. Therefore, the shrink is ¾″ (4 × ³⁄₁₆ = ¾). The shrink is added to the distance to the center of the pipe, and the center bend is marked at that location.

From the table, the multiplier for the distance between bends is 2.61. Therefore, the distance between bends is 10⁷⁄₁₆″ (2.61 × 4 = 10.44″). In order to compensate for the fact that two different benchmarks are being used to make the bend, the outside diameter of the conduit is added to this dimension on each side of the center mark. For ¾″ EMT, the outside diameter is 0.92″ (¹⁵⁄₁₆″). Therefore, the pencil marks for the side bends are placed 11⅜″ (10⁷⁄₁₆ + ¹⁵⁄₁₆ = 11⅜) from the center mark.

To make the first outside bend, the benchmark for the center of a 22½° bend is aligned with the pencil mark for that bend and a 22½° bend is made. The bender can point in either direction.

The next bend is the center bend. The center bend is made in the air. The conduit is rotated 180° in the bender, the benchmark for the center of a 45° bend is aligned with the pencil mark, and a 45° bend is made.

The last bend is the other outside bend. This bend is also made in the air. The conduit is rotated 180° in the bender, the benchmark for the center of a 22½° bend is aligned with the pencil mark, and the 22½° bend is made. Push-through bending is more efficient than other bending methods because the conduit does not have to be turned around in the bender and the offset can be bent with the conduit facing in either direction. This is sometimes a factor when trying to fabricate bends near the end of the conduit.

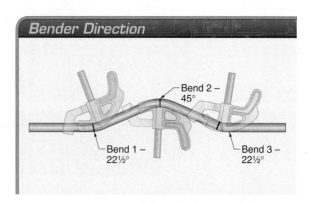

**Bender Direction**

Bend 2 – 45°

Bend 1 – 22½°

Bend 3 – 22½°

## FOUR-BEND SADDLES

Four-bend saddles are used when a conduit run needs to change elevations to clear a large cylindrical obstruction or a rectangular obstruction. Four-bend saddles are two offsets placed in the same plane and in the same run. The four variables in making a four-bend saddle are the same as for making a three-bend saddle.

## Marking the Bend Locations

The center of the obstruction for a four-bend saddle is located in the same manner as the center of an obstruction for a three-bend saddle. The distance to the obstruction, as well as the rise and width, must be measured and the offset bend angle chosen. The shrink must then be calculated and added to the distance to the center of the obstruction.

*Kicks are sometimes used to avoid bending an offset.*

**Edge of Obstruction Method.** There are two methods for placing the pencil marks for a four-bend saddle. In the first method, all bends are made relative to the location of the edge of the obstruction. **See Figure 4-14.** The distance multiplier and shrink constant are the same as for making offset bends. For a 30° bend, the distance multiplier is 2.0 and the shrink constant is ¼. For an obstruction with a 7″ offset rise, the distance between bends is 14″ (2.0 × 7 = 14), and the shrink is 1¾″ (¼ × 7 = 1¾). The width of the obstruction is 12″.

In the first method, mark I is placed at the distance from the beginning of the conduit to the nearest side of the obstruction, plus the shrink. Mark II is placed 14″ back from mark I toward the beginning of the conduit. Mark III is placed 12″ away from mark I toward the far end of the conduit. Mark IV is placed 14″ away from mark III toward the far end of the conduit.

**Center of Obstruction Method.** In the second method, all bends are made relative to the location of the center of the obstruction. The multipliers are the same as in the first method. The distance between bends is 14″ and the shrink is 1¾″. The width of the obstruction is 12″.

In the second method, mark 1 is placed at the distance from the beginning to the center of the obstruction plus the shrink. **See Figure 4-15.** Mark 2 is placed 6″ (half the width) back toward the beginning from mark 1. Mark 3 is placed 14″ back toward the beginning from mark 2. Mark 4 is placed 6″ away from mark 1, toward the far end of the conduit. Mark 5 is placed 14″ from mark 4, toward the far end of the conduit.

This method can be easily adapted to push-through bending by simply adding the take-up, or distance from the end of the bend to the center of the bend, to the dimensions at locations 2 and 4. This will cause a corresponding shift of the marks at locations 3 and 5. The offset can then be bent using the push-through method and the center of bend benchmarks for the respective bends

**Four-Bend Saddle Layout—Edge of Obstruction Method**

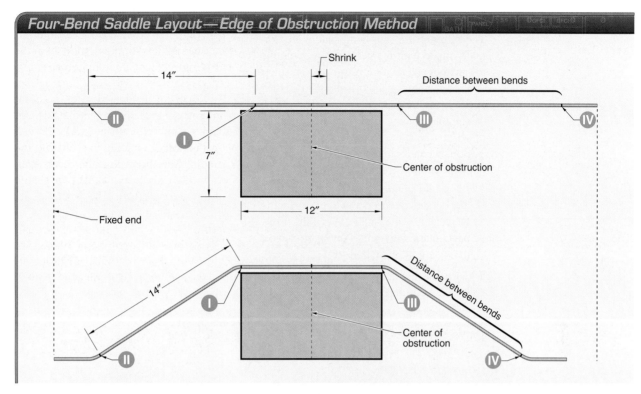

**Figure 4-14.** *When the edge of the obstruction is used as a reference point, shrink is added to the measurement to the edge.*

**Four-Bend Saddle Layout—Center of Obstruction Method**

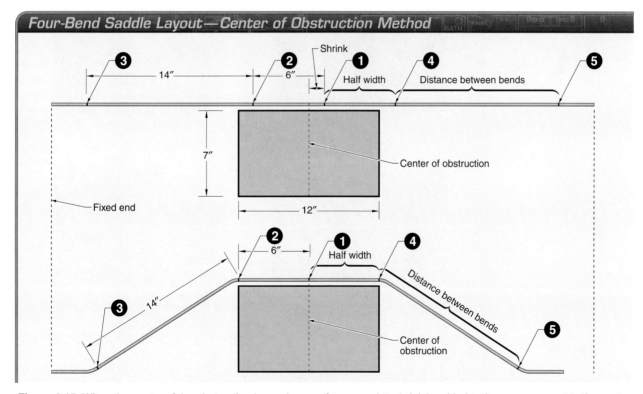

**Figure 4-15.** *When the center of the obstruction is used as a reference point, shrink is added to the measurement to the center.*

## Making the Offset Bends

The offsets for a four-bend saddle are normally made with the bender hook pointing toward the center of the obstruction. The most common angle for the offsets is 30°. However, the offset angles of a four-bend saddle can be made at any angle at which standard offset can be fabricated. **See Figure 4-16.** The offset bends are fabricated with the edge of obstruction method as follows:

1. Place the bender on the conduit with the bender shoe on the ground facing the center of the saddle. Align the arrow benchmark with mark I and fabricate the first bend of the first offset.

2. Turn the bender upside down and rotate the conduit 180°. Align the arrow benchmark with mark II. Carefully align the conduit to prevent a dogleg and start the second bend of the first offset in the air. Complete the bend on the ground.

3. Remove the bender from the conduit and reverse the bender. Place the bender back on the conduit at mark III with the bender shoe facing the center of the saddle. Carefully align the bends with each other and fabricate the first bend of the second offset.

4. Align the arrow benchmark with mark IV. Rotate the conduit 180°, carefully align the conduit with the previous bends, and fabricate the second bend of the second offset.

**Four-Bend Saddles**

1. Align the pencil mark with the arrow benchmark.

2. Turn the bender upside down and make the second bend.

3. Make the first bend of the second offset.

4. Rotate the conduit in the bender and make the last bend.

**Figure 4-16.** *A four-bend saddle consists of two offsets that are equal distances from the center of the obstruction.*

## CORNER OFFSETS

A fabrication similar to a four-bend saddle is a corner offset. A *corner offset* is a bend consisting of two offsets turned at a 90° angle from each other. Corner offsets are used when a conduit run needs to be placed on the adjoining wall of an inside or an outside corner. The 90° angles produce the change in direction that is needed to turn the corner. **See Figure 4-17.**

In the case of a corner offset, the objective is to turn the corner, not to go around an obstruction. The variables are slightly different from the standard four-bend saddle because the center of the obstruction does not have to be found and calculated. Also, it is very likely that each offset will rise different amounts.

## Marking the Bend Locations

First, the amount of offset required to place the conduit on the corner should be found. This is the rise from the location of the conduit run to the edge of the corner. Care must be taken to ensure that the measurement is taken bottom to bottom, as with other offsets. The amount of rise and the chosen offset bend angles are used to calculate the distance between bends for this offset.

Next, the amount of offset necessary to move from the corner to the desired location on the second wall should be found. This is the rise from the desired location of the conduit run on the new wall to the corner. Again, the amount of rise and the chosen offset bend angles are used to calculate the distance between bends for this offset.

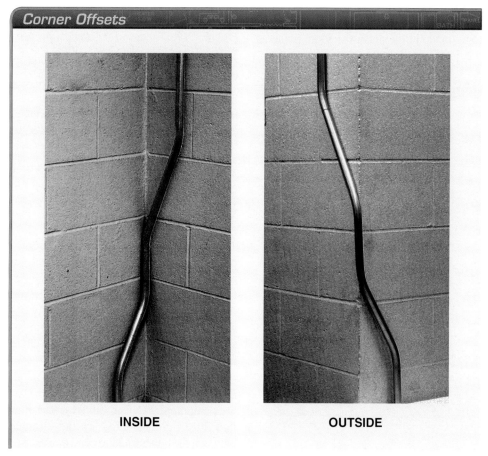

Corner Offsets

INSIDE          OUTSIDE

*Figure 4-17. Corner offsets are used when a conduit run needs to be placed on the adjoining wall of an inside or an outside corner.*

Then these offsets are laid out using the same methods as for any offset bend. The distance between the two offsets is not important as long as the bend looks appropriate and there is sufficient room on the two walls for the offsets to fit properly.

### Making the Offset Bends

The two offsets are then fabricated. The first offset is fabricated just like any other offset. The first bend is made on the ground and the second bend is started in the air and finished on the ground.

For the second offset, the first bend can be made on the ground. However, unlike typical four-bend saddles whose bends are all in the same plane, the second offset needs to be bent in a plane 90° from the first bend. **See Figure 4-18.** An inspection of the field conditions can determine if the bend should be 90° clockwise or counterclockwise from the original offset. The second bend needs to be made in the air in the same plane as the first bend of the second offset.

*Offsets turned 90° from each other*

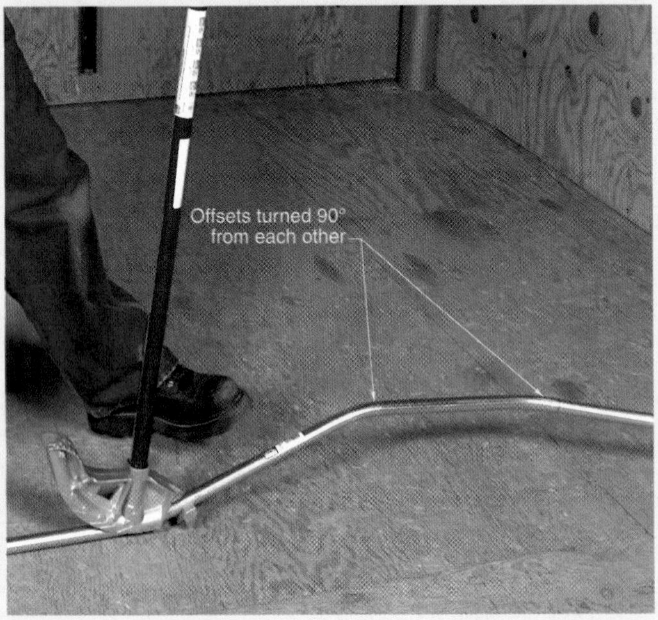

**Figure 4-18.** *The second offset of a corner offset is bent in a plane at 90° from the first bend.*

## COMPOUND 90° BENDS

There are rare instances where a run may call for a compound 90° bend. **See Figure 4-19.** Compound 90° bends are used when an object is obstructing a corner and two bends must be made to turn the corner and get past the obstruction.

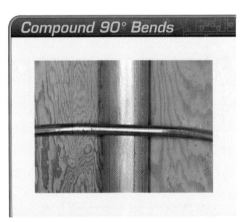

**Figure 4-19.** *Compound 90° bends are used when an object is obstructing a corner and two bends must be made to turn the corner and get past the obstruction.*

**Tech Fact**

An offset bend around a rectangular obstruction is measured to the edge of the obstacle.

### Measured Layout Method

The simplest method of fabricating a compound 90° bend is to use a tape measure or folding rule to measure the distance between bends. The measured layout method is a quick and simple method that does not require any calculations. If using the measured layout method, care must be taken to hold the rule at the proper angles.

**Distance between Bends.** When fabricating compound 90° bends, the two variables that must be considered are the distance between the bends and the bend angles. **See Figure 4-20.** The distance between bends is determined by the size of the obstruction.

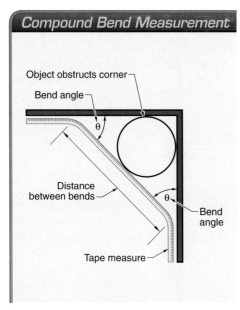

**Compound Bend Measurement**

Object obstructs corner

Bend angle

θ

Distance between bends

θ

Bend angle

Tape measure

**Figure 4-20.** *The two variables that must be considered for compound 90° bends are the distance between the bends and the bend angles.*

The bend angles needed to form the compound 90° bend should also be chosen. Normally, two 45° bends are used to form the bends. However, any combination of angles will work as long as their sum is 90. The procedure to lay out and fabricate a compound 90° bend is as follows:

1. Place the end of the tape or rule on the beginning of the run and bend the tape or rule across the obstruction holding the rule at the chosen angle.

2. Measure the distance to the first bend. Measure the distance to the second bend. Make pencil marks on the conduit at the measured distances.

3. Using the center of the bend benchmark, align the benchmark with the conduit mark and bend a 45° angle (or whichever angle is proper).

4. Reverse the bender and place the handle on the ground. Make certain the bends are kept in the same plane so no doglegs occur and fabricate the second bend.

---

**Tech Fact**

The radius of a bend is measured to the centerline of the conduit.

---

## Calculated Layout Method

A common method of fabricating compound 90° bends is to calculate the distance between the bends based on the type of the obstruction. This method can be used for all compound 90° bends but is usually used on larger conduit or when the obstruction is quite large.

**Distance between Bends.** The first step is to find the distance between the centers of the bends. The multiplier for the calculation for the distance between bends depends on the shape of the obstruction. **See Figure 4-21.** The distance between centers can be calculated as follows:

Round obstruction:
$$distance\ between\ bends =$$
$$diameter \times 2.4$$
$$distance = d \times 2.4$$

Square obstruction:
$$distance\ between\ bends = side \times 3.0$$
$$distance = s \times 3.0$$

Rectangular obstruction:
$$distance\ between\ bends =$$
$$(side\ 1 + side\ 2) \times 1.4$$
$$distance = (s_1 + s_2) \times 1.4$$

where

$distance$ = distance between bends
$d$ = diameter of round obstruction, in inches
$s$ = length of side of square obstruction, in inches
$s_1$ = length of short side of rectangular obstruction, in inches
$s_2$ = length of long side of rectangular obstruction, in inches

For example, if a circular pipe with a diameter of 10″ were in a corner, the distance between bends would be 24″ (10 × 2.4 = 24). If a rectangular obstruction with dimensions of 6″ × 14″ were in a corner, the distance between bends would be 28″ ([6 + 14] × 1.4 = 28).

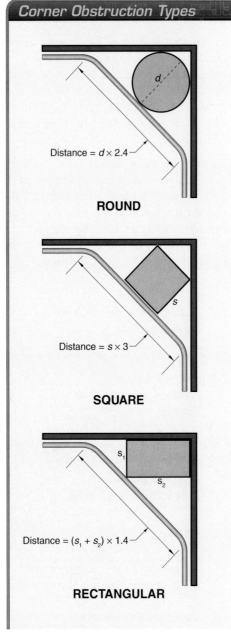

**Corner Obstruction Types**

Distance = $d \times 2.4$

**ROUND**

Distance = $s \times 3$

**SQUARE**

Distance = $(s_1 + s_2) \times 1.4$

**RECTANGULAR**

*Figure 4-21. The distance between bends depends on the size and shape of the obstruction.*

## Application—Compound 90° Bend around a Sprinkler

A compound 90° bend s used to go around a 5″ sprinkler main that is tight to the corner. The origin of the conduit run is 36″ from the corner. **See Figure 4-22.** A 45° bend angle is chosen.

**Marking the Bend Locations.** The first step is to calculate the distance between bends. Since the sprinkler main is round, the distance between bends is calculated as follows:

$$distance = d \times 2.4$$
$$distance = 5 \times 2.4$$
$$distance = \mathbf{12''}$$

The next step is to determine where to place the first bend relative to the origin of the run. Simple trigonometry can be used to calculate the unknown distance from the first bend to the wall. Once this distance is known, the distance to the first bend is easy to determine.

The 12″ distance between bends is the hypotenuse of a triangle. From trigonometry, we know the cosecant of an angle (multiplier) is the ratio of the hypotenuse (distance between bends) to the opposite side (unknown distance). In other words, the unknown distance from the bend to the wall is equal to the distance between bends divided by the offset multiplier. The unknown distance is calculated as follows:

$$unknown\ distance = \frac{distance\ between\ bends}{offset\ multiplier}$$

where

*unknown distance* = distance from first bend to the wall, in inches
*distance between bends* = distance between the two bends in a compound 90° bend, in inches
*offset multiplier* = cosecant of angle

For a 45° angle, the offset multiplier is 1.41. For this example, the distance between the bends is 12″. The unknown distance is calculated as follows:

$$unknown\ distance = \frac{12}{1.41}$$

$$unknown\ distance = \mathbf{8.51''},\ or\ \mathbf{8\frac{1}{2}''}$$

Once the unknown distance is known, the distance to the first bend is easy to calculate. The difference between the distance from the origin to the wall (36″) and the unknown distance (8½″) is the distance to the first bend. **See Figure 4-23.** The distance to the first bend is 27½″ (36 – 8½ = 27½).

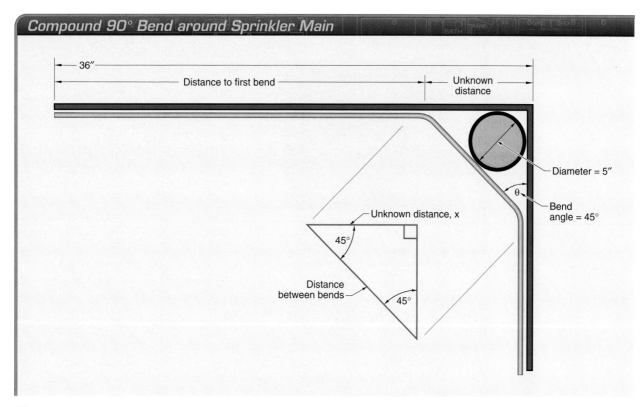

**Figure 4-22.** *A triangle can be used to calculate the location of the pencil marks for compound bends.*

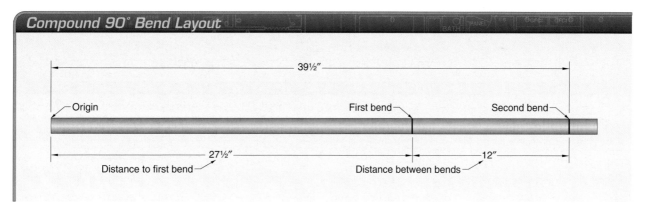

**Figure 4-23.** *After the locations of the bends are calculated, pencil marks are made on the conduit to mark the bends.*

The distance to the first bend and the distance between bends are now both known. The pencil mark for the first bend is placed 27½″ from the end of the conduit. The pencil mark for the second bend is placed 12″ from the first mark (39½″ from the end of the conduit). Both bends are fabricated using the 45° benchmark (rim notch or teardrop) for the center of the bend.

## APPLICATION— THREE-BEND SADDLE AROUND A DRAIN

A conduit must run along an outside wall and cross over a roof drainpipe. **See Figure 4-24.** The conduit is to be run tight along the wall. The rise of the drainpipe is measured at 4¼″ (4.25″). The distance from the last coupling to the edge of the drainpipe is 42⅛″.

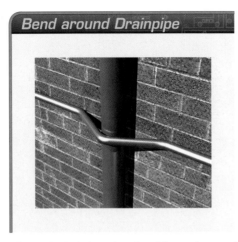

**Figure 4-24.** *A three-bend saddle can be used to bend conduit around a drainpipe.*

Since the drainpipe is circular in cross section, the rise is equal to the width. The distance from the last coupling to the center of the drainpipe is the distance from the last coupling to the edge of the drainpipe plus half the width of the drainpipe. Therefore, the distance to the center of the drainpipe is 44¼″ (42⅛ + 2⅛ = 44¼). **See Figure 4-25.**

A 45° center bend is chosen. Therefore, the side bends will be 22½° each. The shrink constant for a 22½° bend is 0.19″ per inch of rise, and the distance multiplier is 2.5. Therefore, the shrink is ¹³⁄₁₆″ (0.19 × 4.25 = 0.81).

The shrink is added to the distance to the center of the drainpipe to give 45¹⁄₁₆″ (44¼ + ¹³⁄₁₆ = 45¹⁄₁₆). The pencil mark for the center bend is placed this distance from the end of the conduit. The distance between the bends is 10⅝″ (2.5 × 4.25 = 10.625). The two pencil marks for the outside bends are placed this distance from the center mark.

The center bend is fabricated by placing the bender on the conduit, aligning the benchmark for the center of a 45° bend, and bending to 45°. The two 22½° side bends are made in the air with the arrow benchmark aligned with the pencil marks and the bender hook pointed toward the center bend.

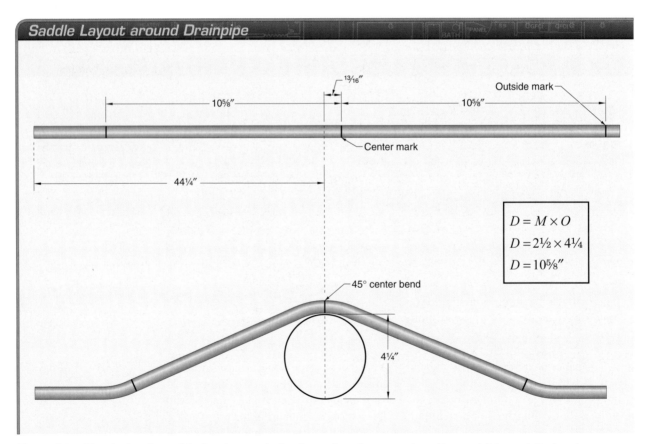

**Figure 4-25.** *After the locations of the bends are calculated, pencil marks are made on the conduit to mark the bends.*

## Finding the Center of a Bend

If the bender does not have a benchmark for the center of the bend, or if a 30° angle is chosen, a new benchmark can be placed on the bender. The procedure to find the center of a bend and create a new benchmark is as follows:

1. Make a pencil mark on the conduit at any convenient location.

2. Align any convenient benchmark with the pencil mark and fabricate a 30° bend on a scrap piece of conduit.

3. Find the center of the bend by extending (with pencil marks) both of the inside edges of the conduit through the bend. The point at which the two marks intersect is the center of the bend.

4. Place the bender back on the conduit and align the original benchmark with the same pencil mark as before. Place a permanent marking at the center of the bend for the 30° bends on the bender at the point where the center of the bend mark touches the shoe.

*1. Place pencil mark.*

*2. Align pencil mark with arrow.*

*3. Find center of bend.*

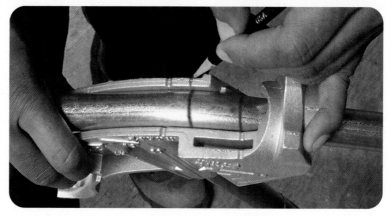

*4. Transfer mark to bender.*

# SUMMARY

- A saddle is a section of conduit consisting of three or four bends that is shaped to bend around an obstacle and then return to its original level.

- A three-bend saddle consists of a center bend with two side bends.

- Three-bend saddles are usually used for relatively small cylindrical obstructions where the obstruction is smaller than the bend radius of the center bend.

- A four-bend saddle consists of two offset bends with a length of straight conduit between the offsets.

- A four-bend saddle has a straight section of conduit between the offsets. This straight section can be made any length to cross over rectangular obstructions.

- The four variables for any saddle bend are the rise of the obstruction, the width of the obstruction, the distance to the obstruction, and the choice of the bend angles.

- Shrink is the product of the shrink constant and the rise.

- The location for the center bend pencil mark is determined by adding the shrink to the measured distance to the obstruction.

- The distance between bends is the product of the rise and the distance multiplier.

- The side bend pencil marks are placed equal distances on either side of the center bend pencil mark.

- A corner offset is a bend consisting of two offsets turned at a 90° angle from each other.

- Corner offsets are used when a conduit run needs to be placed on the adjoining wall of an inside or an outside corner.

- Compound 90° bends are used when an object is obstructing a corner and two bends must be made to turn the corner and get past the obstruction.

# Mechanical and Electric Benders

## CONDUIT BENDING and FABRICATION

## 5

MECHANICAL BENDERS.................................................................78
LAYOUT ...................................................................................81
FABRICATION ..........................................................................88
ELECTRIC BENDERS ..................................................................99
SUMMARY ...............................................................................104

Fabricating conduit using hand benders is a critical skill. However, hand bending has a serious limitation: it is limited to smaller trade sizes. Other tools must be used to bend conduit of larger sizes. In addition, hand bending requires a surface that is hard and somewhat level. It is very difficult to produce accurate, consistent bends on a gravel surface at a job site.

Two common types of benders that have been developed to bend larger conduit are mechanical and electric benders. Many of the same skills an electrician masters with hand benders are useful for mechanical and electric benders.

The push-through method is used with mechanical and electric benders because it is very difficult to reverse the conduit for subsequent bends as is commonly done with hand benders. This method requires that a bender be charted to obtain information, such as take-up, gain, setback, and travel, that is used to pre-position bends on the conduit before it is placed in the bender.

### OBJECTIVES

1. Explain the major differences between mechanical and electric benders.
2. List the components of both mechanical and electric benders.
3. Explain the operating principles of both mechanical and electric benders.
4. Describe how to chart a bender.
5. Describe how to use the push-through method to fabricate offsets and saddles.
6. Explain how to fabricate a kick with the measured rise method and with the multiplier method.
7. Describe how to use a no-dog tool with a mechanical or electric bender.

## MECHANICAL BENDERS

A *mechanical bender* is a type of bender that employs a lever arm and ratcheting mechanism to provide a mechanical advantage when fabricating conduit bends. **See Figure 5-1.** Mechanical benders are typically known as Chicago benders. Mechanical benders can easily be modified to raise the bender higher above the ground to make it easier to work with. They have been used for many years and have proven to be accurate and dependable tools.

Mechanical benders are relatively simple to operate and will bend rigid conduit ranging in size from ½″ to 1½″. They are also relatively lightweight, are easily transportable, can be set up in many types of environments, do not require electricity, and are extraordinarily rugged. It is quite possible for one electrician to transport a mechanical bender into the middle of a field, complete the required fabrications, and return the bender back to the trailer.

---

### Tech Fact

Mechanical benders are typically used to bend conduit up to 2″ trade size. A bender typically has a shoe for each conduit size, along with a hook or collar that holds the conduit against the shoe.

---

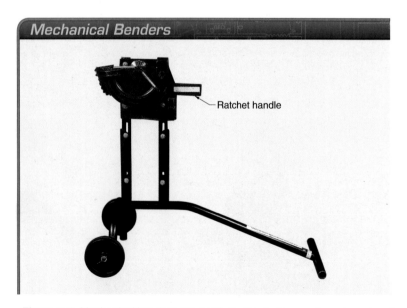

**Mechanical Benders**

Ratchet handle

*Figure 5-1. Mechanical benders are relatively simple to operate and can bend rigid conduit ranging in size from ½″ to 2″.*

## Mechanical Bender Components

Mechanical benders have some components in common with hand benders. A hand bender has a shoe that is used to form the bend. A mechanical bender has several shoes that are used for different sizes of conduit. **See Figure 5-2.** A hand bender has a hook that holds the conduit against the shoe. A mechanical bender has a hook or removable collar that holds the conduit against the shoe. In addition, a mechanical bender has a rear conduit support or roller that carries the conduit as it is drawn through the bender.

Hand benders have markings that indicate the amount of the bend. Although a mechanical bender has a degree scale, the scale is generally not accurate enough for precise conduit work. Instead, the degree of bend will usually be determined by using the distance that the conduit travels into the bender as an indicator of bend angle. Since the bender can predictably create the desired angle of bend, the conduit can be laid out using the multiplier method before it is inserted into the bender. By using the multiplier method, travel, and a push-through technique, an electrician can fabricate offsets, kicks, and 90° bends in conduits with a high degree of accuracy and efficiency.

## Bending Aids

Because of the design of the mechanical benders, it can be somewhat challenging to bend a kick, offset, or three- or four-bend saddle. It is impossible to turn the bender upside-down to check alignment as is done with hand benders. A common method used to achieve accurate alignment and avoid doglegs is to use a "no-dog," or "anti-dog" tool.

A no-dog is a level vial mounted in a thumbscrew clamp. **See Figure 5-3.** Once the conduit has been properly chucked up in the bender, but before the conduit is actually bent, the no-dog is clamped on the end of the conduit in the level position. The bend is then fabricated to the correct angle, the bender is released, and the conduit is moved forward to the mark for the second bend.

**Mechanical Bender Components**

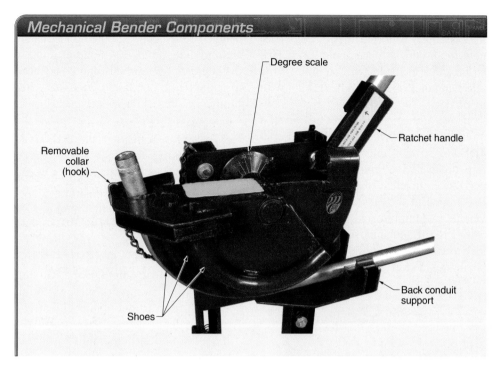

Degree scale

Ratchet handle

Removable collar (hook)

Back conduit support

Shoes

*Figure 5-2.* Mechanical benders can have several shoes and corresponding back supports that are used for different sizes of conduit.

**No-Dogs**

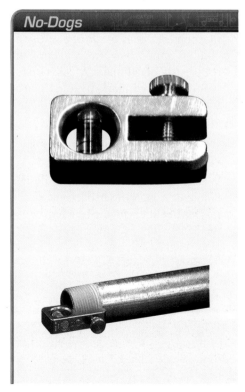

*Figure 5-3.* A common method to achieve accurate alignment and avoid doglegs is the use of a "no-dog" or "anti-dog" tool.

Large power centers often require long conduit runs with many complex bends.

Before the next bend is completed, the conduit is rotated 180° in the bender until the no-dog is once again level. This ensures that the bends will be in the same plane and the conduit will be straight with no doglegs.

If one of these tools is not available, a no-dog can easily be made from a scrap piece of strut, a conduit clamp, and a torpedo level. **See Figure 5-4.** With a torpedo level attached to the strut, the strut is clamped to the conduit with the level vial set at plumb. Once the first bend is completed, the conduit is rotated 180°, and the strut and conduit are plumbed before beginning the next bend.

**Figure 5-4.** A no-dog can easily be made from a scrap piece of strut, a conduit clamp, and a torpedo level.

## Mechanical Bender Operation

A unique component on a mechanical bender is the ratcheting mechanism, which includes a pawl release. Releasing the pawl can be dangerous because the weight of the conduit in front of the hook or to the side of the bender can cause the conduit to shift

suddenly. Also, releasing the pawl frees the handle of the bender, allowing the handle and the shoe to move without restraint. In the interest of safety, care should be taken to support the conduit if necessary. **See Figure 5-5.** In general, a mechanical bender is operated as follows:

1. Mark the conduit at the desired location(s).

2. Release the pawl and rotate the shoe to its fully retracted position. Select the correct shoe for the conduit size and place the conduit in the bender. Attach the front hook (if necessary), engage the pawl, and lift or lower the front handle to engage the ratchet. Align the pencil mark on the conduit with the corresponding benchmark on the bender. Allow the weight of the handle to hold the conduit captive in the bender.

3. If bending an offset or a saddle, place a no-dog on the end of the conduit and carefully rotate the conduit until the no-dog is level.

4. Use a convenient nonmoving part of the bender, such as the back of the conduit support or the center of the roller, as a benchmark. Measure back from the benchmark by the desired travel distance and place a pencil mark on the conduit.

5. Pull down on the handle to start the bend.

6. When the conduit has moved the desired amount through the bender, the pencil mark will align with the benchmark and the bend will be complete.

---

### Tech Fact

According to the NEC®, metal raceways, cable armor, and other metal enclosures for conductors shall be metallically joined together into a continuous electric conductor and shall be connected to all boxes, fittings, and cabinets so as to provide effective electrical continuity. Unless specifically permitted elsewhere in the Code, raceways and cable assemblies shall be mechanically secured to boxes, fittings, cabinets, and other enclosures.

## Mechanical Bender Operation

1. Mark the conduit at the desired location(s).

2. Align the pencil mark on the conduit with the corresponding benchmark on the bender.

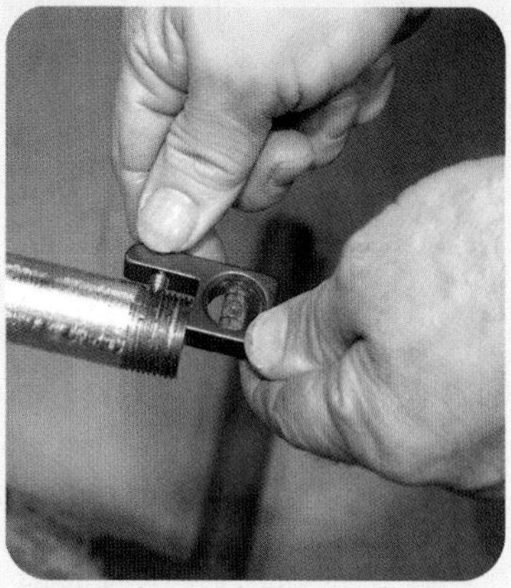

3. Place a no-dog on the end of the conduit and carefully rotate the conduit until the no-dog is level.

4. Measure back from the benchmark by the desired travel distance and place a pencil mark on the conduit.

5. Pull down on the handle to start the bend.

6. The bend is complete when the pencil mark aligns with the benchmark.

**Figure 5-5.** *In a mechanical bender, the conduit moves through the bender during bending. Travel marks on the conduit indicate when the bend is complete.*

## LAYOUT

Many of the same skills the electrician mastered using hand benders are also useful with mechanical benders. The layout procedures learned for hand bending also apply to bending with mechanical benders. Two differences between hand bending and bending with a mechanical bender are the use of the push-through method and the types of benchmarks used for pre-positioning bends.

While it is a common practice to turn a hand bender around and bend the second bend in the opposite direction, it is fairly unusual to turn the conduit around in a mechanical bender. When using a hand bender, the conduit is generally aligned with the arrow, notch, or star benchmarks. No such benchmarks exist on a mechanical bender. Using the front of the hook on a mechanical bender as the benchmark for making a 90° bend is similar to using the arrow on the hand bender, but offsets and kicks are made by aligning the conduit markings with the center-of-bend benchmark on the shoe. It is also possible to mark the conduit so that the markings are lined up with the back of the support at the rear of the bender. Mechanical benders normally use the push-through method.

## Push-Through Method

A *push-through method* is any procedure for bending conduit in which the conduit is fed through a bender and not turned around end for end during the bending process. Instead, multiple bends are made from one end to the other as the conduit is moved through the bender. This method is sometimes called the "amount of travel" method and usually refers to bends made with mechanical benders. However, push-through methods can be used with any type of bender.

A drawback to the push-through method is that the benchmarks for centers of the various standard bends are rarely found on any type of bender shoe. Sample bends must be fabricated and the centers of the bends must be transferred to the bender shoe as permanent markings.

## Charting a Bender

The first step in using the push-through method of bending conduit is charting the bender. The procedure used to chart a bender is also used to determine the take-up, gain, setback, radius adjustment, and amount of travel. This step is vitally important because the accuracy of all future bends made with the bender depends upon having accurate values in order to do the layout and fabrication work. Each bender should have a chart. **See Figure 5-6.** A chart for a bender is created as follows:

1. Make a pencil mark on a piece of conduit 6″ from the end (mark 1). Measure the length of the conduit piece. If the conduit is a full length, it may not be convenient to use the full length for measurements. In this case, place another mark (mark 4) 40″ down from the end. Mark 1 is to be aligned with the hook at the front of the

shoe and used to determine take-up. The length of the conduit or mark 4 will be used to determine the gain.

2. Place the conduit in the bender and carefully align mark 1 with the front of the hook, with the 6″ length extending past the hook. Engage the pawl and allow the handle of the bender to maintain pressure on the conduit. Recheck to verify that there is exactly 6″ of conduit ahead of the hook.

3. Place a mark a set distance back from the conduit support or another fixed point at the rear of the bender (mark 2) and record the distance.

4. Advance the bender until a 90° bend is formed. Check this bend carefully, using a level as the bend is completed. It is important to fabricate a bend that is exactly 90°.

5. Once the bend is formed, make a pencil mark (mark 3) at the same conduit support or fixed point at the rear of the bender that was used to make mark 2.

6. Measure the distance between mark 2 and mark 3. Subtract this distance from the original distance to mark 2. This difference is the travel. Remove the conduit from the bender and verify that the degree of bend is exactly 90° by checking it with a level or a square. Record the take-up, gain, setback, radius adjustment, and 90° travel.

Creating a chart for a bender may require work, but a written log or chart that displays the amount of travel for each shoe on each type of conduit can be permanently kept on each bender. This can be a great help when fabricating conduit on an unfamiliar bender or coming back to a bender after spending time on another task. The information should be written down and kept for future reference. When a similar bender is encountered, these measurements can be used to start using the bender, thus saving valuable time.

**Take-up.** Take-up is the distance from the back of the bend to the pencil mark when making a 90° bend. Take-up is used to determine where to position the pencil mark used to fabricate accurate 90° bends and stubs. For hand benders, take-up is typically given on the bender itself. For mechanical benders, take-up may be stamped or painted on the bender. If take-up is not known for a bender, it can easily be determined by measuring the distance from the back of the leg of the bend to the pencil mark (mark 1) used to chart the bender. This is the take-up. **See Figure 5-7.**

**Gain.** Gain is the shortcut formed by bending on a curve instead of at a right angle. The gain of a particular bender shoe may be marked on the bender. If the gain is not available, it can be determined by measuring the stub length (from the back of the bend to the end of the stub) and leg length (from the back of the bend to the end of the conduit or from the back of the bend to mark 4) and subtracting the original length. **See Figure 5-8.** The result is the gain. For example, a piece of conduit with a length of 40″ is bent into a stub of 14″ and a leg of 29″. The sum of the stub and leg lengths is 43″ (14 + 29 = 43), and the gain is 3″ (43 − 40 = 3).

**Setback.** Setback, or shrink back, is the difference between take-up and gain. Setback is used to determine the location of the back of a 90° bend relative to the take-up mark. Once the take-up mark has been made, the setback can be added to the take-up distance and the position of the back of the bend will be known. If an offset needs to be fabricated a specific distance from the back of the stub, measurements and layout marks for the offset can be made based on the setback mark. This allows both 90° bends and offsets to be laid out on the same piece of conduit prior to putting the conduit in the bender.

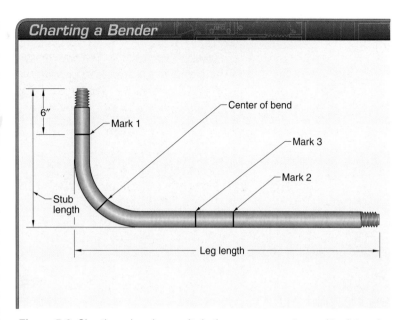

*Figure 5-6.* Charting a bender results in the measurements used to determine take-up, gain, setback, radius adjustment, and travel.

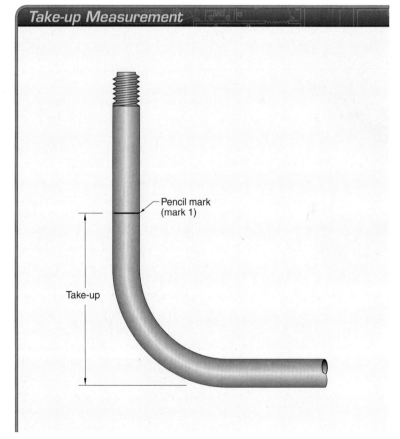

*Figure 5-7.* The take-up is the distance from the back of the bend to the pencil mark.

---

*T*ech Fact

According to the NEC®, bends shall be made so that the conduit will not be damaged and so that the internal diameter of the conduit will not be effectively reduced. The radius of the curve of any field bend to the centerline of the conduit shall not be less than indicated in the Code.

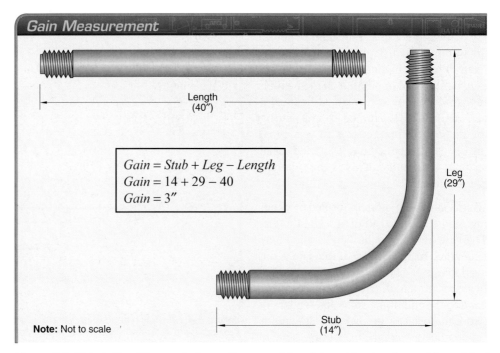

**Figure 5-8.** *Gain is the difference between the sum of the stub and leg lengths and the original conduit length.*

**Radius Adjustment.** With the push-through method of making bends, a benchmark for the center of a bend is often used to bend offsets and saddles. If a bend is centered on a pencil mark that is used to locate the edge of an obstruction, the center of the bend will be aligned with the edge of the obstruction. This results in a poor fit. A radius adjustment can be made to move the bend farther away from the edge and place the end of the bend at the edge of the obstruction. The radius adjustment is then subtracted from the sum of the shrink and the distance to the obstruction.

To determine the radius adjustment, a bend of the desired angle is made, the center is located, and the end-of-bend point is located where the conduit is once again straight. **See Figure 5-9.** The end-of-bend point can be found be placing a straightedge or another piece of conduit of the same size against the outside of the conduit in which the bend has been made. A piece of paper is inserted between the conduit and the straightedge and slid toward the end of the bend until it can go no further. The conduit is marked at this point and the distance to the center of the bend is measured.

This distance is the radius adjustment and should be included as a value on the chart for the bender.

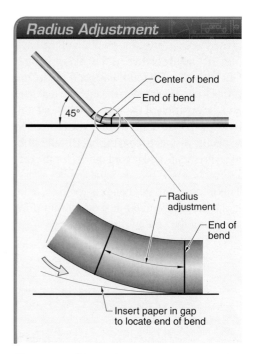

**Figure 5-9.** *The radius adjustment is used to move the center of a bend away from the edge of an obstruction.*

As a rule, the radius adjustment for a 45° bend can be used with other bend angles with acceptable results. However, if maximum accuracy must be achieved, the radius adjustment for the specific angle of bend should be determined.

**Travel.** With a mechanical bender, the linear movement of the conduit past the conduit support or roller is directly proportional to the angle of the bend. This linear movement is called the travel. **See Figure 5-10.** *Note:* This value is different for different benders and different types of conduit.

---

## Tech Fact

According to the NEC®, conduit shall be installed as a complete system and shall be securely fastened in place and supported.

---

**Travel**

Amount of travel for 30° bend

Pencil marks for 30° bend

Pencil marks move by amount of travel

**Figure 5-10.** *The linear movement of the conduit through the follow bar is directly proportional to the angle of the bend.*

If the amount of travel is not known, it can be determined by charting the bender. The 90° travel is the difference between the original and final distances from the conduit support to the end of the conduit, or the difference between the original and final distances from the conduit support to mark 2. This is the amount the conduit travels through the bender for a 90° bend for that type and size of conduit.

For example, fabricating ¾″ rigid on one particular bender takes 8″ of linear movement to make a 90° bend. For this situation, the conduit is placed in the bender, and a pencil mark is placed 8″ down from a benchmark such as the center of the roller or the back of the conduit support. As the bend is fabricated, the pencil mark is drawn closer to the benchmark. The bend is 90° when the pencil mark aligns with the benchmark and no pressure is being applied to the bender handle. Using the push-through method, the bender produces accurate bends in every kind of situation.

Because of the linear nature of this method, the amount of travel for any angle can easily be calculated. A 45° bend is half of a 90° bend. Therefore, the amount of travel of a 45° bend is half the amount of travel of a 90° bend. A 30° bend is one-third of a 90° bend. Therefore, the amount of travel of a 30° bend is one-third the amount of travel of a 90° bend. The exact amount of travel can be calculated for any angle as follows:

$$travel = (90°\, travel) \times \frac{angle}{90}$$

where

*travel* = amount of travel for the desired angle, in inches
*90° travel* = amount of travel for a 90° bend, in inches
*angle* = required bend angle, in degrees

For example, with a 90° amount of travel of 8″ and a required 30° bend angle, the required amount of travel will be 2.67″. This can be calculated as follows:

$$travel = (90° \, travel) \times \frac{angle}{90}$$

$$travel = 8 \times \frac{30}{90}$$

$$travel = \mathbf{2.67''}$$

**Travel per Degree.** An alternative to determining the travel for a 90° bend is to calculate the travel per degree of bend. For any angle, the amount of travel is the product of the angle and the travel per degree. The travel per degree value is constant for one type of conduit on one shoe. This means there is a different travel per degree value for every type of conduit and for every shoe. The travel is calculated as follows:

$$travel = travel \, per \, degree \times angle$$

where

*travel* = amount of travel for the desired angle, in inches

*travel per degree* = amount of travel for each degree of bend, in inches

*angle* = required bend angle, in degrees

For this method, the travel per degree is the amount of travel for a 90° bend divided by 90. For example, if the amount of travel for 90° is 8″, the travel per degree is 0.0889 (8 ÷ 90 = 0.0889). The amount of travel for a 30° bend with the same type of conduit and the same type of shoe is calculated as follows:

$$travel = travel \, per \, degree \times angle$$

$$travel = 0.0889 \times 30$$

$$travel = \mathbf{2.67''}$$

## Pre-Positioning Bends

The procedures used to pre-position bends with hand benders are also used with mechanical benders. These procedures include calculating the distance between bends and the shrink.

**Distance between Bends.** There are many types of fabrications that consist of more than one bend. For any bends that are made to rise over an obstruction, the distance between the bends is the product of the distance multiplier and the rise, often called the offset rise. For offset bends, the distance multiplier is the cotangent of the bend angle and is typically given in tables. Since a four-bend saddle consists of two offsets, the same multiplier is used. Adjustments are sometimes made to these multipliers to account for the use of different benchmarks for different bends.

The most common way to fabricate three-bend saddles with the push-through method is to use a distance multiplier of 3.0. For example, a three-bend saddle with a 45° center bend has to rise over a 4¾″ (4.75″) obstruction. The push-through method will be used. The distance between bends is calculated as follows:

$$D = M \times O$$

where

*D* = distance between bends, in inches

*M* = distance multiplier

*O* = offset rise, in inches

$$D = M \times O$$

$$D = 3.0 \times 4.75$$

$$D = \mathbf{14.25''} \text{ or } \mathbf{14\frac{1}{4}''}$$

For hand-bent offsets, the distance multiplier is 2.6 when a 22½° offset bend is used. For hand-bent three-bend saddles, the distance multiplier is 2.5 when a 22½° side bend is used. The arrow benchmark is used for the side bends, and the center-of-bend benchmark is used for the center bend. The multiplier is changed from 2.6 to 2.5 to compensate for using different benchmarks for the side bends and center bend.

For the push-through method of making three-bend saddles, the distance multiplier is 3.0 when a 45° center bend is used. The center-of-bend benchmark is used for each of the bends. Each bend angle has a different center-of-bend benchmark. The distance multiplier of 3.0 is typically used instead of the standard multiplier of 2.6 to compensate for using different benchmarks. However, using this multiplier often results in the distance between bends being too large and the center bend being too high above the obstruction. A more accurate but more complex method is to use the standard distance multiplier of 2.6. **See Figure 5-11.**

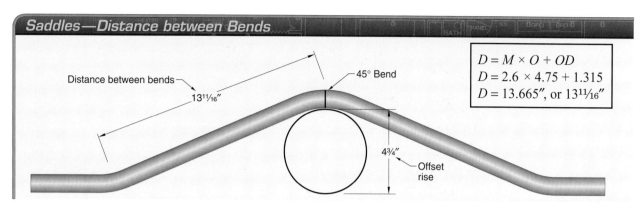

**Figure 5-11.** *The distance between bends is calculated from the offset rise and the conduit OD.*

However, the distance between bends needs to be increased by the OD of the conduit. For example, 1″ rigid conduit is used to fabricate a three-bend saddle with a 45° center bend that has to rise over a 4¾″ (4.75″) obstruction. The OD of 1″ rigid is 1.315″. The push-through method will be used. The distance between bends is calculated as follows:

$$D = M \times O + OD$$

where

$D$ = distance between bends, in inches

$M$ = distance multiplier

$O$ = offset rise, in inches

$OD$ = outside diameter

$$D = M \times O + OD$$
$$D = 2.6 \times 4.75 + 1.315$$
$$D = \textbf{13.665″, or } \textbf{13}^{\textbf{11}}\!/\!_{\textbf{16}}″$$

**Shrink.** Any time conduit is bent to rise over an obstruction, the total run the conduit can cover is reduced. This shrink is the product of the shrink constant and the offset rise. **See Figure 5-12.** For example, a four-bend saddle is used to rise over an obstruction. The height of the obstruction is 6½″ (6.5″) and a 30° bend angle is chosen. The shrink constant for a 30° bend is ¼. The shrink for the first offset of the saddle is calculated as follows:

$$S = C \times O$$

where

$S$ = shrink, in inches

$C$ = shrink constant

$O$ = offset rise, in inches

$$S = C \times O$$
$$S = \tfrac{1}{4} \times 6.5$$
$$S = \textbf{1.625″, or } \textbf{1}^{\textbf{5}}\!/\!_{\textbf{8}}″$$

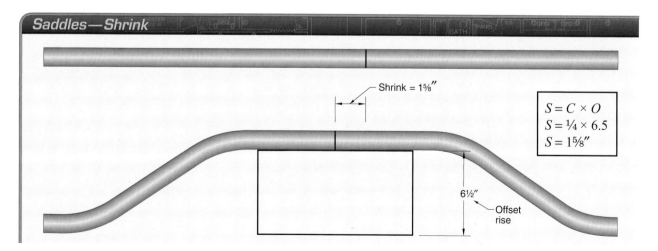

**Figure 5-12.** *Shrink is the product of the shrink constant and the rise.*

## FABRICATION

After conduit bends are laid out, mechanical benders can be used to fabricate the bends. A lever arm is used to advance the conduit through the bender. All the standard bends that can be made with hand benders can also be used with mechanical benders.

### 90° Bends

For all benders, the take-up must be known in order to fabricate a 90° bend to the desired length. The take-up may be stamped or written on the bender. If the take-up is not known, it must be measured on a 90° bend made in a piece of conduit. After the take-up is determined, a 90° bend is laid out in a similar way to a hand bend. The travel is used to produce the correct bend angle. **See Figure 5-13.** A 90° bend is fabricated as follows:

1. Measure the height of the required stub. Subtract the take-up from the measurement and place a pencil mark on the conduit.

2. Place the conduit in the bender and align the pencil mark with the same benchmark that was used to measure take-up. The benchmark will typically be the front of the hook.

3. Measure back from the conduit support, the center of the roller, or another stationary point at the back of the bender by the 90° travel distance and place a pencil mark on the conduit.

4. Fabricate the 90° bend. The bend is complete when the pencil mark advances so that it is even with the benchmark when no pressure is being applied to the handle. Verify the bend with a level or protractor.

For example, 1″ rigid conduit is to be used to make a 90° bend. The take-up for the bender is 8″ and the amount of travel for a 90° bend is 9⅞″. The 90° bend is fabricated as follows:

1. The stub height is measured as 18″. Place a pencil mark 10″ (18 − 8 = 10) from the end of the conduit.

2. Place the conduit in the bender with the pencil mark aligned with the front of the hook.

3. Place a travel pencil mark 9⅞″ back from the travel benchmark to indicate the amount of travel.

4. Fabricate the bend by allowing the bender to draw the conduit through the bender until the travel pencil mark advances to the travel benchmark. This indicates that a 90° bend has been completed. The stub should extend 18″ from the back of the bend.

### Kicks

Using mechanical benders to fabricate kicks is fairly straightforward since the same techniques used with hand benders apply. Kicks can be fabricated with the multiplier method and with the measured rise method.

**Multiplier Method for Kicks.** The amount of travel can be used to bend accurate angles on a mechanical bender. The multiplier method requires benchmarks to be made on the bender for the center of the desired angle of bend. The conduit is laid out so that the center-of-bend pencil mark on the conduit will coincide with the center-of-bend benchmark on the bender.

The measurements for a kick can be laid out after the 90° bend is fabricated, but it is more efficient to do the entire layout first in order to save time. The key to doing this is to use the setback in order to determine where the center-of-bend mark for the kick will have to be placed. By using the take-up, gain, shrink, and setback, the conduit can be cut, threaded (if needed), and then bent, thus reducing the overall amount of labor needed.

For example, a ¾″ rigid metal conduit is to be fabricated with a 12″ stub, an 8″ kick bent at 30°, and an overall run length of 44″ from the back of the stub to the end of the leg. The bender take-up is 8⁹⁄₁₆″, the gain is 3¼″, and the 90° travel is 7³⁄₁₆″. The setback is 5⁵⁄₁₆″ (8⁹⁄₁₆ − 3¼ = 5⁵⁄₁₆) and the travel for a 30° bend is 2⅜″ (⅓ × 7³⁄₁₆ = 2⅜).

## Mechanical Benders—90° Bends

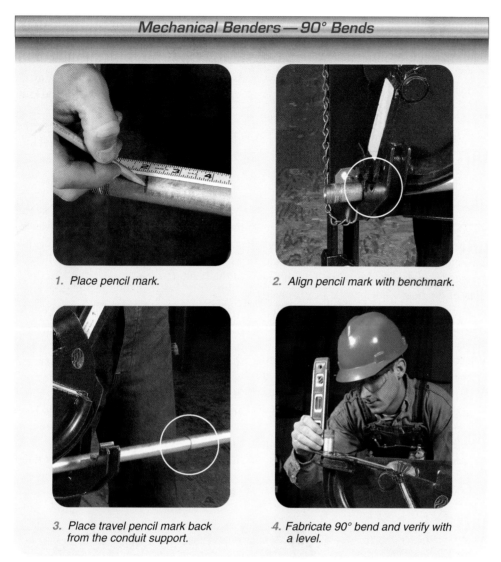

1. Place pencil mark.

2. Align pencil mark with benchmark.

3. Place travel pencil mark back from the conduit support.

4. Fabricate 90° bend and verify with a level.

**Figure 5-13.** *The amount of travel can easily be used to fabricate 90° bends with a mechanical bender.*

The conduit is laid out by making a pencil mark (mark 1) on the conduit at 3⁷⁄₁₆″ (12 − 8⁹⁄₁₆). **See Figure 5-14.** Another pencil mark (mark 2) is placed an additional 5⁵⁄₁₆″ down the conduit for the setback. Mark 2 represents the side or back of the stub after the bend is complete. This mark is used only for layout, not for bending.

For a 30° bend, the shrink constant is ¼ and the distance multiplier is 2.0. Therefore, the shrink is 2″ (¼ × 8 = 2). The calculated distance between the center of the stub bend and the kick bend is 16″ (2.0 × 8 = 16). An adjustment to the distance to the center of the kick of half the OD is made. This

is to compensate for the fact that mark 2 coincides with the back of the stub, not the center. The OD of ¾″ rigid is 1.050. Half the OD is 0.525 (1.050 ÷ 2 = 0.525). Therefore, the distance between bends is 16½″ (16 + 0.525 = 16.525). A pencil mark (mark 3) is placed an additional 16½″ down from mark 2 for the kick bend. This mark is the center-of-bend mark for the kick.

The total length of conduit needed is the sum of the lengths of the stub and leg, plus the shrink, and minus the gain. The stub is 12″, the leg is 44″, the shrink is 2″, and the gain is 3¼″. Therefore, the total length of conduit is 54¾″ (12 + 44 + 2 − 3¼ = 54¾). The conduit should be cut and threaded at 54¾″.

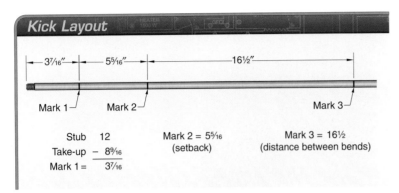

**Kick Layout**

—3⁷⁄₁₆″—|—5⁵⁄₁₆″—|————16½″————

Mark 1      Mark 2                    Mark 3

| Stub | 12 | Mark 2 = 5⁵⁄₁₆ | Mark 3 = 16½ |
| Take-up | – 8⁹⁄₁₆ | (setback) | (distance between bends) |
| Mark 1 = | 3⁷⁄₁₆ | | |

*Figure 5-14. Setback is used to lay out a kick.*

The 90° bend is fabricated in the normal manner by inserting the conduit into the bender and aligning mark 1 with the front of the hook. The bender is advanced so that the conduit is held captive and a pencil mark placed for the travel distance of 7³⁄₁₆″ back from the travel benchmark. The conduit is advanced until the 90° bend for the stub is completed. This is indicated when the travel mark is aligned with the travel benchmark. The kick is fabricated as follows:

1. Release the bender and slide the conduit forward until mark 3 is aligned with the benchmark for the center of the 30° bend. **See Figure 5-15.**

2. Advance the bender to hold the conduit captive and then rotate the conduit 90° and level the stub.

3. Mark the 30° travel of 2³⁄₈″ from the travel benchmark and fabricate the kick using the travel mark. If desired, verify the height of the kick by placing the conduit on a flat, level surface, leveling the stub, and measuring from the surface to the bottom of the conduit.

When advancing the bender, it will generally be necessary to go slightly beyond the desired value of the kick in order to compensate for the springback that occurs when pressure is removed from the handle of the bender.

**Measured Rise Method for Kicks.** For the kick made with the measured rise method, the 90° bend is fabricated in the normal manner. The desired distance between the bends is chosen and the conduit is pushed through the bender by that amount. The conduit is rotated 90°, as with the multiplier method. The amount the stub rises is measured during fabrication of the kick. When the stub has risen the desired distance, the bend is complete. **See Figure 5-16.**

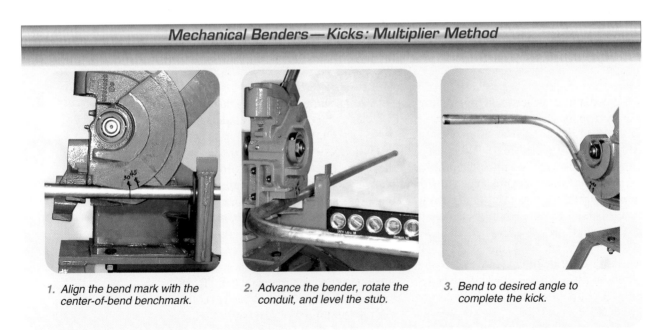

*Mechanical Benders—Kicks: Multiplier Method*

1. *Align the bend mark with the center-of-bend benchmark.*

2. *Advance the bender, rotate the conduit, and level the stub.*

3. *Bend to desired angle to complete the kick.*

*Figure 5-15. Kicks consist of a 90° bend and an angled bend to raise the stub.*

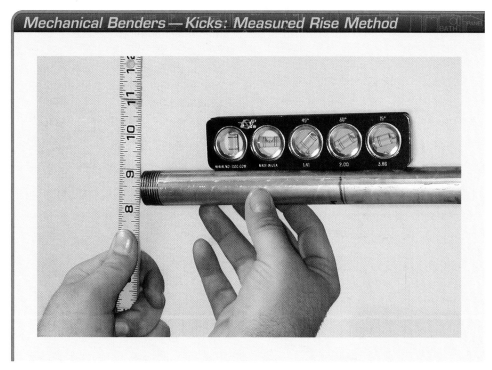

**Figure 5-16.** *The measured rise method for kicks uses a rule or tape measure to measure the rise of the kick. The measured rise is the distance the stub rises above the original conduit run. The rise can be verified after the conduit is removed from the bender.*

For the measured rise method of fabricating kicks, the exact angle of bend is not a concern as there is some leeway in how far the conduit needs to be moved. It must be kept in mind that if the conduit is only advanced a short distance, the kick can end up being nearly 90° in some cases. This would be unsightly and cause wire pulling to be more difficult. On the other hand, if the conduit is advanced too far, the kick may end up interfering with other conduits or equipment.

For example, an 8″ kick is desired in a piece of conduit, along with a 12″ stub. The conduit is ¾″ RMC, and the take-up of the bender is 9¾″. The 90° stub is fabricated by marking the conduit at 2¼″ (12 − 9¾ = 2¼) from the end, aligning the mark with the front of the hook, and bending it to 90°. The kick is started by moving the conduit forward about 10″, rotating the conduit 90°, and measuring the height of the stub above the floor.

If the measurement from the bottom of the conduit to the floor is 13¼″, the bender will have to be advanced until the bottom of the conduit is at 21¼″ (13¼ + 8 = 21¼) above the floor when the bend is complete and pressure on the handle has been released. If desired, the height of the kick can be verified by placing the conduit on a flat level surface, leveling the stub, and measuring from the surface to the bottom of the stub.

In this scenario, the bender is presumed to be level from side to side, and the 90° stub is presumed to be exactly plumb after it has been bent (but still in the bender). If it is found that the bender is tilted to one side, it will be necessary to apply an equal amount of correction to the stub when leveling it in order to make the kick.

## Offsets

With a mechanical bender, it is difficult to bend an offset in the same manner as with a hand bender because it is inconvenient to release the conduit from the bender when fabricating several bends.

Therefore, the push-through method is normally used to make offsets with mechanical benders.

The push-through method allows both bends of an offset to be fabricated by starting at one end and working toward the other. The most common method of fabricating offsets uses the center-of-bend benchmark as the reference point when making bends. New benchmarks must be placed on the bender shoe for the center of each desired bend.

**Offset Layout.** Laying out an offset for the push-through method is similar to laying out an offset in hand bending, but there is a significant difference. In hand bending, offsets are usually made using the arrow mark on the bender as the benchmark. Since the arrow typically coincides very closely with the end of the bend, no adjustment is needed for the position of the marks on the conduit. However, when bending conduit using the push-through method and the center-of-bend benchmark, the layout method must be modified slightly in order to create an offset where the bend is completed before the obstruction is reached.

**Fabricating Offset Bends.** The height of the offset rise and the distance to the edge of the obstruction must be measured before beginning. The bend angle must be chosen and the shrink, distance between bends, radius adjustment, and amount of travel for the chosen bend angle all need to be determined. **See Figure 5-17.** An offset is fabricated as follows:

1. Add the shrink to the distance to the edge of the offset and then subtract the radius adjustment. Place a pencil mark at that point. Measure back by the distance between bends and place another pencil mark.

2. Place the conduit in the bender and align the first pencil mark with the center-of-bend benchmark for the chosen angle. Slowly advance the bender until the conduit is held firmly, but do not start the bend.

3. Attach a no-dog and level it on the end of the conduit. Place a travel pencil mark back from the roller or the support to mark the amount of travel.

4. Complete the first bend by advancing the conduit through the bender until the travel mark lines up with the roller or support. If desired, verify the bend angle with a protractor or level.

5. Release the bender and advance the conduit until the second bend mark aligns with the center-of-bend benchmark for the chosen angle. Rotate the conduit 180°. Slowly advance the bender until the conduit is held firmly, but do not start the bend. Use the no-dog to verify that the conduit has been rotated exactly 180°. If necessary, rotate the conduit with pliers or vise grips.

6. Place a travel pencil mark back from the roller to mark the amount of travel. Complete the second bend by advancing the conduit through the bender until the travel mark lines up with the benchmark. If desired, verify the bend angle with a protractor or level.

---

**Tech Fact**

According to the NEC®, electrical equipment shall be installed in a neat and workmanlike manner.

---

For example, ¾″ rigid conduit is to be used for an offset. For this bender, the take-up is 8⁹⁄₁₆″, the gain is 3¼″, and the amount of travel for a 90° bend is 7³⁄₁₆″. The offset rise is 14″ and the distance to the edge of the obstruction is 52″. A 30° bend angle is chosen. The shrink constant is ¼ and the distance multiplier is 2.0. The shrink is 3½″ (¼ × 14 = 3½) and the distance between bends is 28″ (2.0 × 14 = 28). The radius adjustment is 2¾″. Since a 30° angle is ⅓ of a 90° angle, the amount of travel is 2⅜″ (⅓ × 7.1875 = 2.395). The offset is fabricated as follows:

1. Place a pencil mark 52¾″ (52 + 3½ − 2¾ = 52¾) from the end of the conduit. Measure back 28″ toward the beginning of the conduit from the pencil mark and place the pencil mark for the first bend.

2. Place the conduit in the bender and align the pencil mark for the first bend with the benchmark for a 30° angle. Slowly advance the bender until the conduit is held firmly, but do not start the bend.

3. Attach a no-dog to level the end of the conduit and level it. Place a travel pencil mark 2⅜″ back from the back of the conduit support to mark the amount of travel.

4. Complete the first bend by advancing the conduit through the bender until the travel mark moves 2⅜″ and lines up with the back of the conduit support.

5. Release the bender and advance the conduit until the second bend mark aligns with the center-of-bend benchmark for the chosen angle. Rotate the conduit 180°. Slowly advance the bender until the conduit is held firmly, but do not start the bend. Use the no-dog level to verify that the conduit has been rotated exactly 180°.

6. Place a travel pencil mark 2⅜″ back from the roller to mark the amount of travel. Complete the second bend by advancing the conduit through the bender until the travel mark lines up with the back of the conduit support.

## Mechanical Benders—Offsets

1. Place pencil marks for the two bends.

2. Align the first pencil mark with the center-of-bend benchmark.

3. Place travel pencil mark and attach no-dog.

4. Use the amount of travel to fabricate the first bend. If desired, verify with protractor.

5. Rotate the conduit and align with the center-of-bend benchmark. Place another travel pencil mark.

6. Use the amount of travel to fabricate the second bend. If desired, verify with a protractor.

**Figure 5-17.** An offset bend is made with two equal bends.

## Saddles

With a mechanical bender, it is impractical to bend a saddle in the same manner as with a hand bender. While it is possible to remove the conduit from the bender and reverse the direction of bend, maximum efficiency will be achieved if all of the bends are made with the conduit facing the same direction. Therefore, the push-through method is normally used to make saddles with mechanical benders.

**Three-Bend Saddles.** The push-through method allows all three bends to be fabricated by starting at one end and working toward the other. With the push-through method for three-bend saddles, the conduit is laid out the same as with a hand bender. The rise is measured, and the shrink is calculated and added to the center mark. The angles are chosen and the distance between bends is calculated. When using a mechanical bender, all of the bends are aligned with the center-of-bend benchmarks on the bender shoe.

As with a hand bender, the same basic bends are used: a 45° bend in the middle and two 22½° bends on the ends. With a hand bender, the center bend is made first. With the push-through method, whether using a hand, mechanical, or electric bender, a side 22½° bend is made first, the center 45° bend second, and the other 22½° bend third.

When using the push-through method to bend a three-bend saddle, it is necessary to compensate for the fact that the center-of-bend benchmark for the two side bends is not in the same location as the benchmark for the center bend. For an offset, the distance between bends for a 22½° bend would be calculated by simply multiplying the desired rise by a multiplier of 2.61. However, when making a three-bend saddle, it is necessary to add the outer diameter of the conduit to the distance calculated. This is because the distance between the 45° and 22½° benchmarks on the bender shoe is approximately the diameter of the conduit. Therefore, to calculate the center-of-bend distance for a three-bend saddle, the following formula is used:

$$Distance = distance\ multiplier \times offset\ rise + OD$$

$$D = M \times O + OD$$

For example, a three-bend saddle made from 1″ EMT needs to cross a drain line running down a wall. **See Figure 5-18.** The drain line extends 3⅜″ (3.375″) out from the wall and the center is located 34″ from the end of the previous conduit run. For the 22½° side bends, the shrink constant is 3/16 and the distance multiplier is 2.6. The amount of travel is 8½″ for a 90° bend. The OD of 1″ EMT is 1.163″.

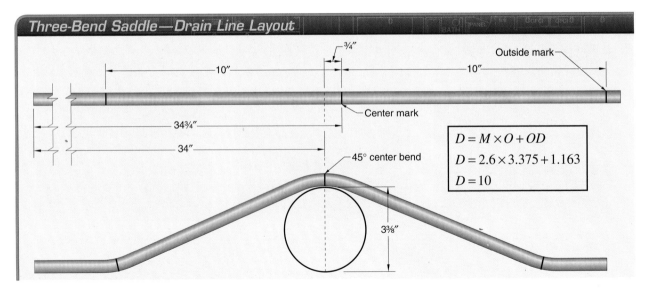

**Figure 5-18.** A three-bend saddle can be used to cross over a drain line.

The shrink is ⅝″ (³⁄₁₆ × 3⅜ = 0.63), and the distance between bends is 10″ (2.6 × 3.375 + 1.163 = 9.94). Since 22½° is ¼ of 90°, the amount of travel for a 22½° bend is 2⅛″ (¼ × 8½ = 2⅛), and the amount of travel for the center 45° bend is 4¼″ (½ × 8½ = 4¼). **See Figure 5-19.** The saddle is fabricated as follows:

1. Place a pencil mark 34⅝″ (34 + ⅝ = 34⅝) from the end. This is the center of the center bend. Place two more pencil marks 10″ on either side of the center mark.

2. Place the conduit in the bender and align the first pencil mark with the benchmark for the center of a 22½° bend. Advance the bender to hold the conduit firmly. Place a no-dog on the end of the conduit and level it. Place a travel pencil mark 2⅛″ back from the travel benchmark. Complete the first 22½° side bend by advancing the conduit in the bender until the travel mark aligns with the travel benchmark.

3. Slide the conduit through the bender to align the second pencil mark with the benchmark for the center of the 45° bend. Advance the bender to hold the conduit, rotate the conduit 180°, and again level the conduit with the no-dog. Place a travel pencil mark 4¼″ back from the travel benchmark. (Keep in mind that this bend requires twice the travel of the other two.) Complete the center 45° bend by advancing the conduit in the bender until the travel mark aligns with the benchmark. If desired, verify the bend angle with a protractor or level.

4. Slide the conduit through the bender to align the third pencil mark with the benchmark for the center of the second 22½° bend. Advance the bender to hold the conduit, rotate the conduit 180° for the last bend and use the no-dog to level the conduit. Place a travel pencil mark 2⅛″ back from the travel benchmark. Complete the last 22½° side bend. If desired, verify the bend angle with a protractor or level.

**Four-Bend Saddles.** The push-through method is normally used to create four-bend saddles when using a mechanical or electric bender. As with a hand bender, the four-bend saddle is simply two offsets with a straight section of conduit between them. With a hand bender, a bend near the center is made first, then a side bend is made to finish the first offset. The other offset is then made with the bender facing in the opposite direction. However, with a mechanical bender, all bends are made starting at one end and moving to the other.

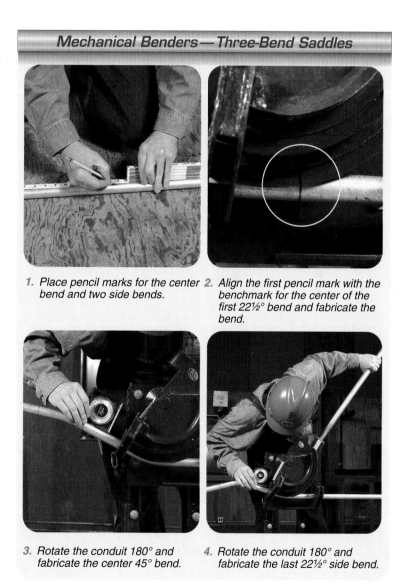

### Mechanical Benders—Three-Bend Saddles

1. Place pencil marks for the center bend and two side bends.

2. Align the first pencil mark with the benchmark for the center of the first 22½° bend and fabricate the bend.

3. Rotate the conduit 180° and fabricate the center 45° bend.

4. Rotate the conduit 180° and fabricate the last 22½° side bend.

**Figure 5-19.** The basic bends of a three-bend saddle are a 45° center bend with 22½° side bends.

With the push-through method for four-bend saddles, the conduit is laid out the same way for any type of bender. All of the bends are aligned with the center-of-bend benchmarks on the bender shoe, and the conduit is pushed through the bender as each bend is made. The rise is measured, the angles are chosen, and the distance between bends is calculated. The shrink is calculated and added to the center mark. The width of the obstruction is measured and used as the distance between the two offsets, along with a radius adjustment made when using the center-of-bend benchmarks.

For example, a four-bend saddle needs to be fabricated using 2″ rigid conduit. For the bender being used, take-up is 15¾″, gain is 6⅞″, and the amount of travel for a 90° bend is 15″. The offset rise is 14″ and the distance to the edge of the obstruction is 48″. The obstruction is 24″ wide. Half the obstruction width is 12″. **See Figure 5-20.**

A 30° bend angle is chosen. The shrink constant is ¼ and the distance multiplier is 2.0. The shrink is 3½″ (¼ × 14 = 3½) and the distance between bends is 28″ (2.0 × 14 = 28). The radius adjustment is 4″. Since a 30° angle is ⅓ of a 90° angle, the amount of travel is 5″ (⅓ × 15 = 5).

The four-bend saddle is laid out by placing a pencil mark (mark 1) at 63½″ (48 + 12 + 3.5 = 63½) from the end. This mark will end up at the center of the obstruction. The distance from mark 1 to mark 2 and from mark 1 to mark 4 is half the width of the obstruction (12″) plus the radius adjustment (4″). Therefore distance from mark 1 to mark 2 and from mark 1 to mark 4 is 16″ (12 + 4 = 16). Mark 2 is placed 16″ to the left of mark 1, and mark 4 is placed 16″ to the right of mark 1.

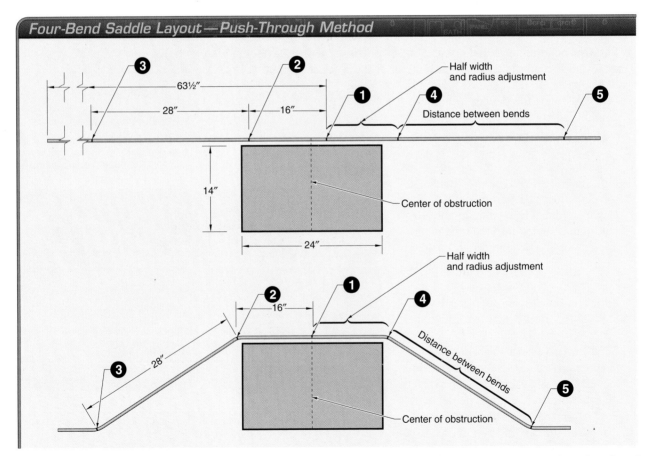

*Figure 5-20. For the push-through method, a radius adjustment is used to move the center of the bends away from the edge of the obstruction.*

The distance from mark 2 to mark 3 and from mark 4 to mark 5 is the center-to-center bend distance for both of the offsets. The center-to-center bend distance is the product of the offset multiplier (2.0) and the offset rise (14″). Therefore the center-to-center bend distance for the offset is 28″ (2.0 × 14). Mark 3 is placed 28″ to the left of mark 2, and mark 5 is placed 28″ to the right of mark 4. Marks 2 through 5 should be extended all the way around the conduit, and mark 1 can be erased to help limit confusion during the bending process.

At this point, it should be noted that mark 5 is within about 10″ of the far end of the conduit. This will create a problem when fabricating the last bend because there may not be enough conduit at the end to properly mark and use the travel. Since all benchmarks are for the center of a bend, this saddle can be started with the first bend at mark 5, and then pushed into the bender so that the last bend will be made at mark 3. **See Figure 5-21.** A four-bend saddle is fabricated as follows:

1. Place the conduit in the bender and align mark 5 with the center-of-bend benchmark for a 30° angle. Slowly advance the bender until the conduit is held firmly, but do not start the bend.

2. Attach a no-dog and level the end of the conduit. Place a travel pencil mark 5″ back from the travel benchmark to mark the amount of travel.

3. Complete the first bend by advancing the conduit through the bender until the travel mark lines up with the benchmark.

4. Release the conduit from the bender, rotate it 180°, and advance it until the second bend mark (mark 4) aligns with the center-of-bend benchmark for a 30° bend. Slowly advance the bender until the conduit is held firmly, but do not start the bend. Use the no-dog level to verify that the conduit has been rotated exactly 180°. If necessary, turn the conduit with pliers or a pipe wrench.

5. Place a travel pencil mark 5″ back from the travel benchmark to mark the amount of travel. Complete the second bend.

6. Release the conduit from the bender, but this time do not rotate the conduit. Move it until the third bend mark (mark 2) aligns with the center-of-bend benchmark for a 30° bend. Slowly advance the bender until the conduit is held firmly, and verify that the no-dog is level.

7. Place a travel pencil mark 5″ back from the travel benchmark and complete the third bend.

8. Release the conduit from the bender, rotate it 180°, and advance it until the fourth bend mark (mark 3) aligns with the center-of-bend benchmark for a 30° bend. There will also be a problem fabricating the last bend if the bender is on the floor and the bends are being made in the vertical plane because the conduit will run into the floor. It will be necessary to elevate the bender or hang the conduit over the edge of a landing or stairwell in order to complete the last bend. If using an electric bender, it may be possible to tilt the bender 90° and make the bends in the horizontal plane.

9. Slowly advance the bender until the conduit is held firmly, and once again use a no-dog to level the conduit. Mark the travel for the fourth bend and complete the bend.

**Tech Fact**

According to the NEC®, where the equipment grounding conductor consists of a raceway, it shall be installed in accordance with the applicable provisions of the Code, using fittings for joints and terminations approved for use with the type raceway or cable used. All connections, joints, and fittings shall be made tight using suitable tools.

## Mechanical Benders—Four-Bend Saddles

1. Place the conduit in the bender and align the first pencil mark with the center-of-bend benchmark.

2. Attach a no-dog level to the end of the conduit.

3. Advance the conduit through the bender until the travel mark lines up with the conduit support.

4. Release the conduit from the bender and rotate it 180°.

5. Advance the conduit through the bender.

6. Move the conduit and verify that the no-dog is level.

7. Complete the third bend by advancing the conduit through the bender until the travel mark lines up with the conduit support.

8. If necessary, elevate the bender in order to complete the last bend.

9. Complete the fourth bend by advancing the conduit through the bender until the travel mark lines up with the conduit support.

**Figure 5-21.** A four-bend saddle consists of two offsets equally spaced from the center of the obstruction. The most common bends are 30°.

## ELECTRIC BENDERS

An *electric bender* is a type of bender that uses an electric motor to rotate a series of shoes that fabricate conduit bends. **See Figure 5-22.** Electric benders can also be used in the field as long as an adequate power source is available. However, electric benders are very heavy and generally require two or more people to safely move them.

**Electric Benders**

*Figure 5-22. An electric bender is a type of bender that uses an electric motor to rotate a series of shoes that fabricate conduit bends.*

Electric benders should be protected from the elements at all times. They are very expensive and represent a major investment. The main advantages of electric benders over other types of benders are speed and accuracy. Once the operating instructions are mastered, the electrician can produce very accurate bends in a very short amount of time. Electric benders are ideal for running large racks of conduit when accurate repeatable bends are required. The multiplier method may be used on all offsets and kicks bent with an electric bender, regardless of the angle or the size and type of conduit.

## Electric Bender Components

An electric bender has an electric motor, a gearbox that rotates a series of shoes, and a hook or collar to hold the conduit in the shoe. **See Figure 5-23.** Electric benders also have a support bar or a series of rollers designed to hold the conduit in place as the bend is being made. The area of the bender carrying the shoe can be rotated just over 90° to help fabricate certain types of bends. Also, some electric benders have a mechanism that allows the electrician to zero the bender, choose the type and size of the conduit to be bent, and sometimes dial in an angle value.

## Electric Bender Operation

Different models of mechanical benders are very similar to each other. However, electric benders can vary significantly from model to model, even models from the same manufacturer. Because of these differences, it generally takes time to learn to use each model. However, bend layout is very similar to that of mechanical benders. Any bend that can be manufactured with a mechanical bender can be manufactured with an electric bender. Electric benders can do the job more quickly and with less physical labor. Consequently, electric benders are well-suited to jobs where there is a lot of large conduit to be installed.

An instruction or operation manual comes with each machine. Whenever possible, an electrician should read this manual before using the bender for the first time. The manual contains important safety information in addition to the operating instructions.

---

**Tech Fact**

The length of the conduit run and the method that will be used to fish in the wiring must be considered. If using a traditional fish tape, a rule of thumb is to limit the length of conduit between pulling points to less than 100′, unless the fish tape is longer and the pipe run is fairly straight.

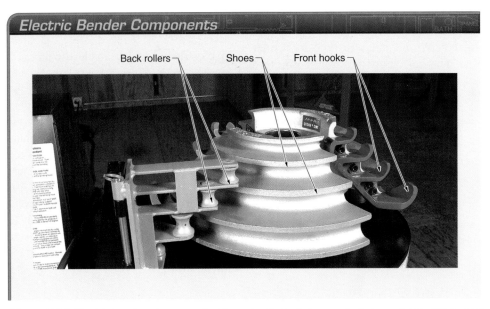

Back rollers — Shoes — Front hooks —

**Figure 5-23.** *Electric benders have an electric motor that rotates a set of shoes, back rollers, and front hooks or collars that hold the conduit in the shoes.*

Since electric benders can fabricate accurate bend angles, the various multiplier methods for making bends work very well. **See Figure 5-24.** In general, an electric bender is operated as follows:

1. Mark the conduit at the desired location(s). Place the conduit in the bender and align the pencil mark on the conduit with the corresponding benchmark on the bender. Advance the bender until the conduit is just snug in the bender. The conduit should be able to be rotated with a pair of pliers or a wrench.

2. If bending an offset or a saddle, place a no-dog on the end of the conduit and carefully rotate the conduit until the no-dog is level.

3. Advance the bender to the desired angle of bend, using the handheld switch or the control pad.

The table on the bender and the degree scale can be used to determine the angle of bend on simple electric benders, but the completed angle must be checked with a level or protractor. The values given by a manufacturer are only approximate and cannot be relied upon without field verification.

Travel can also be used to determine the bend angle, in the same manner as with a mechanical bender.

There are a number of electric benders which have additional features that allow the bend angle to be precisely fabricated by means of the electronic circuitry within the bender. Once the bender is set up for a given size and type of conduit, very accurate bends can be made very quickly. There are a number of different ways to establish the angle of bend with these benders, and the operation manual for the bender will describe the setup procedure.

## Kicks

Virtually any bend that can be manufactured with a mechanical bender can be manufactured with an electric bender. The procedures for laying out bends for the different kinds of benders are the same from one kind of bender to another. For example, kicks can easily be fabricated using the measured rise method with either a mechanical or electric bender. The only significant difference is the use of some type of electric controller instead of the ratchet used with a mechanical bender.

## Electric Bender Operation

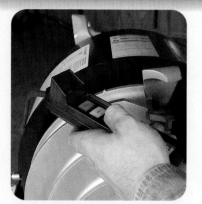

1. Align the pencil mark with the appropriate benchmark.

2. Place a no-dog on the end of the conduit.

3. Use the control pad to fabricate the desired bend.

**Figure 5-24.** *Electric benders have a control pad to operate the bender, but use the same layout procedures as mechanical benders.*

**Measured Rise Method.** The measured rise method can be used to fabricate kicks on an electric bender. The 90° bend is fabricated in the normal manner. For the kick, the measured rise method simply measures the amount the stub rises during kick fabrication. When the stub has risen the desired distance, the bend is complete.

For example, a 12″ kick is desired in a piece of conduit, along with a 13″ stub. The conduit is ¾″ rigid, and the take up of the bender is 8¼″. **See Figure 5-25.** A kick is fabricated by the measured rise method as follows:

1. Fabricate the 90° stub by marking the conduit at 4¾″ (13 – 8¼ = 4¾) from the end, aligning the mark with the front of the hook, and bending it to 90°.

2. Start the kick by moving the conduit forward about 24″, rotating the conduit 90°, checking with a level, and measuring the height above the floor.

3. The amount of rise is measured as the rise above the height above the floor. Finish the kick by advancing the bender until the conduit rises by the desired amount of kick, in this case 12″.

## Electric Benders—Kicks: Measured Rise Method

1. Align the first pencil mark with the benchmark and fabricate a 90° bend.

2. Advance the conduit in the bender, rotate it 90°, and verify that the stub is level.

3. When the stub has moved by the desired amount, the bend is complete.

**Figure 5-25.** *The measured rise method for kicks measures the amount of travel the stub makes during kick fabrication.*

For hand bending, some simple calculations are used to determine where to place the pencil marks on the conduit to ensure that the bends are made in the correct places. The calculations for both shrink and the distance between bends are simplifications of more complicated calculations.

The simplified formulas work well for hand bending. However, when bending large conduit, the simplifications no longer give good results. Larger conduit is bent with a larger bend radius and the exact shrink and gain need to be determined. Electricians that bend large conduit will find that the following method of pre-positioning gives much better results.

### Shrink

The simplified method used for hand bending calculates shrink as the product of the offset rise and the shrink constant. The offset rise is measured in the field. The shrink constant depends on the angle of the bend and is available in tables.

These calculations are not as accurate as desired when bending large conduit. Using the simplified calculations with large conduit, the calculated shrink results in conduit that is too long after the bends are completed. With large conduit, the bend radius must be taken into account as well as the offset rise and the bend angle as determined previously. The bend radius is usually given in the bender manual. When the bend radius is taken into account, the shrink is calculated as follows:

$$S = C \times O + \frac{2 \times R \times \theta \times \pi}{180} - 4 \times R \times C$$

where

$S$ = shrink, in inches
$C$ = shrink constant
$O$ = offset rise, in inches
$\theta$ = bend angle, in degrees
$R$ = bend radius, in inches

For example, a 45° offset is to be bent in 4″ conduit over a 14″ obstruction. From the shrink constant table, the shrink constant is 0.41. The offset rise is 14″. The shrink amount from the simplified method is calculated as follows:

$$S = C \times O$$
$$S = 14 \times 0.41$$
$$S = \mathbf{5.74″}, \text{ or } \mathbf{5\frac{3}{4}″}$$

In order to calculate the actual amount of shrink, the bend radius must also be known. For this situation, a typical bend radius is 20″. Note that this amount can vary quite a bit with different benders. The shrink amount from the complete method is calculated as follows:

$$S = C \times O + \frac{2 \times R \times \theta \times \pi}{180} - 4 \times R \times C$$
$$S = 0.41 \times 14 + \frac{2 \times 20 \times 45 \times 3.14}{180} - 4 \times 20 \times 0.41$$
$$S = 5.74 + 31.40 - 32.80$$
$$S = \mathbf{4.34″}, \text{ or } \mathbf{4\frac{5}{16}″}$$

The actual shrink for this offset bend is 4⁵⁄₁₆″. The shrink calculated by the simplified method is 5¾″. This means that a piece of conduit that was laid out with the simplified method will end up a little over 1⅜″ too long when it is ready to install. The difference is greater for bends at larger angles.

| Distance Multiplier and Shrink Constant | | |
|---|---|---|
| Bend Angle, θ | Distance Multiplier | Shrink Constant |
| 5° | 11.4 | 0.044 |
| 10° | 5.76 | 0.087 |
| 15° | 3.86 | 0.13 |
| 22½° | 2.61 | 0.20 |
| 30° | 2.00 | 0.27 |
| 45° | 1.41 | 0.41 |

### Distance between Bends

In addition to shrink adjustments, the calculation for the distance between bends needs an adjustment when bending large conduit. The simplified method for hand bending calculates the distance between bends as the product of the offset rise and the distance multiplier. The distance multiplier is the cosecant of the angle and is available in tables. The distance between bends is calculated as follows:

$$D = M \times O$$

where

$D$ = distance between bends, in inches
$M$ = distance multiplier
$O$ = offset rise, in inches

With large conduit, the bend radius must be taken into account as well as the offset rise and the bend angle as determined previously. When the bend radius is taken into account, the distance between bends is calculated as follows:

$$D = M \times O + \frac{R \times \theta \times \pi}{180} - 2 \times R \times tan\left(\frac{\theta}{2}\right)$$

where

$D$ = distance between bends, in inches
$M$ = distance multiplier
$O$ = offset rise, in inches
$R$ = bend radius, in inches
$\theta$ = bend angle, in degrees

For example, a 45° offset is to be bent in 4″ conduit over a 14″ obstruction. Therefore, the offset rise is 14″. From the distance-between-bends table, the distance multiplier is 1.41. The distance between bends from the simplified method is calculated as follows:

$$D = M \times O$$
$$D = 1.41 \times 14$$
$$D = \textbf{19.74″, or 19¾″}$$

In order to calculate the actual distance between bends, the bend radius must also be known. For this situation, a typical bend radius is 20″. Note that this amount can vary quite a bit with different benders. The distance between bends from the complete method is calculated as follows:

$$D = M \times O + \frac{R \times \theta \times \pi}{180} - 2 \times R \times tan\left(\frac{\theta}{2}\right)$$

$$D = 1.41 \times 14 + \frac{20 \times 45 \times 3.14}{180} -$$
$$2 \times 20 \times tan\left(\frac{45}{2}\right)$$

$$D = 19.74 + 15.70 - 40 \times 0.414$$
$$D = 19.74 + 15.70 - 16.56$$
$$D = \textbf{18.88″, or 18⅞″}$$

The actual distance between bends for this offset is 18⅞″. The distance between bends calculated by the simplified method is 19¾″. This means that the distance between bends will end up about ⅞″ too long and will result in a poor fit around the obstruction.

## SUMMARY

- Mechanical benders use a lever arm and ratcheting mechanism to provide a mechanical advantage when fabricating conduit bends.

- Electric benders have an electric motor, a gearbox that rotates a series of shoes, and a hook or collar to hold the conduit in the shoe.

- A push-through method is any procedure for bending conduit in which the conduit is not turned around end-for-end during the bend.

- With mechanical and electric benders, offsets, saddles, and kicks are normally fabricated using a push-through method.

- The amount of travel of the conduit through the bender is related to the angle of the bend.

- A bender chart gives the take-up, gain, setback, radius adjustment, and amount of travel for each bender shoe.

- A no-dog is a bending tool used to ensure that all bends are made in the same plane.

- Bend layout procedures are the same for mechanical and electric benders.

# Hydraulic Benders

## CONDUIT BENDING and FABRICATION

**6**

HYDRAULIC PRINCIPLES........................................................106
HYDRAULIC BENDERS..........................................................106
LAYOUT AND FABRICATION ................................................109
SUMMARY ...........................................................................114

**B**enders that use hydraulic systems are available from a number of manufacturers. All hydraulic benders have a hydraulic cylinder, a ram, and a shoe, and a frame or yoke to hold them in the proper position. On some benders, the conduit is bent against rollers, while on others, a follow bar and a pipe collar are used to help form the bend.

All bender hydraulic systems include the following components: pumps to build system pressure, valves to control system pressure and fluid flow, hoses to carry the hydraulic fluid from the pump to the cylinder, a piston and ram to drive the shoe or follow bar, and a fluid reservoir to keep the volume of hydraulic fluid constant. Hydraulic systems operate at high pressure, and care must be taken when operating a hydraulic bender because of the risk of serious injury.

Two general categories of hydraulic benders are one-shot benders and multiple-shot benders. One-shot benders can fabricate a 90° bend in one operation. Multiple-shot benders require two or more operations to fabricate a 90° bend. Regardless of the details of the bender operation, layout for conduit bent using hydraulic benders is generally the same as for conduit bent using hand benders.

### OBJECTIVES

1. Explain why it is necessary to perform a daily inspection of a hydraulic bender before use.
2. Explain how hydraulic fluid flows through a hydraulic system.
3. List the major components of a hydraulic bender.
4. Explain how ram travel can be used to fabricate repeatable bend angles.
5. Explain how to fabricate a 90° bend or an offset with a hydraulic bender.

## HYDRAULIC PRINCIPLES

A hydraulic bender uses a hydraulic system to provide the power required to bend large conduit. Hydraulic systems operate at high pressure, and care must be taken when operating a hydraulic bender because of the risk of serious injury.

### Pascal's Law

All hydraulic systems are based on Pascal's law. Pascal's law states that pressure in an enclosed vessel of fluid is transmitted to all sides of the vessel. When that pressure is applied over an area, such as the end of a piston, a force is transferred to that area. Useful work can be done if the piston is attached to a tool, such as the ram of a bender.

### Hydraulic System Safety

The pressure and force developed in hydraulic benders can be very dangerous. A hydraulic bender should not be used unless the operation and risks are clearly understood. If an accident occurs, serious injury can result.

The pressure required to bend large, heavy-walled conduit is very high. Contact with escaping fluid under extremely high pressure can actually inject the hot fluid under a worker's skin, creating serious medical problems.

Also, when a hydraulic system operates for an extended period of time, the hydraulic fluid can become dangerously hot. If a leak occurs, the electrician can be seriously burned. All hoses, attachments, connectors, and fittings should be inspected daily before use. Hoses and fittings should be kept clean. Dirt or other contaminants in the hydraulic system will ruin the O-rings and seals in the hose couplings and cylinders.

### Hydraulic System Components

Different types of hydraulic benders are all slightly different from each other. **See Figure 6-1.** While varying greatly in type and style, all bender hydraulic systems include the following common components:

- pump to build system pressure
- valves to control system pressure and fluid flow
- hoses to carry the hydraulic fluid from the pump to the cylinder
- piston and ram to drive the shoe or follow bar
- fluid reservoir to keep the volume of hydraulic fluid constant

When the pump is engaged, the system is pressurized. When a valve is opened, the hydraulic fluid moves through the hoses and pressurizes the cylinder. If the hydraulic pressure is high enough to do the work required, the ram begins to move as the hydraulic fluid fills the cylinder.

As the hydraulic pressure moves the ram, additional hydraulic fluid is pumped into the cylinder, maintaining system pressure. A fluid reservoir is required to maintain the proper fluid volume. While the control valve is closed, the cylinder remains pressurized. With hydraulic benders, the valve is usually designed to open automatically when the system is shut down.

## HYDRAULIC BENDERS

Hydraulic benders are available from a number of manufacturers and are used to fabricate 90° bends, offsets, and saddles in large conduit. The principles developed for hand bending are also used with hydraulic benders. Using hydraulic benders is the most practical method for bending the larger trade sizes of conduit.

---

*Tech Fact*

A rubber O-ring can be used to keep track of the amount of ram travel. A rubber O-ring can be placed over the travel rod and positioned to correspond with the desired travel distance. When the O-ring touches the top of the bender, the conduit will be bent to the desired angle.

## Hydraulic System Components

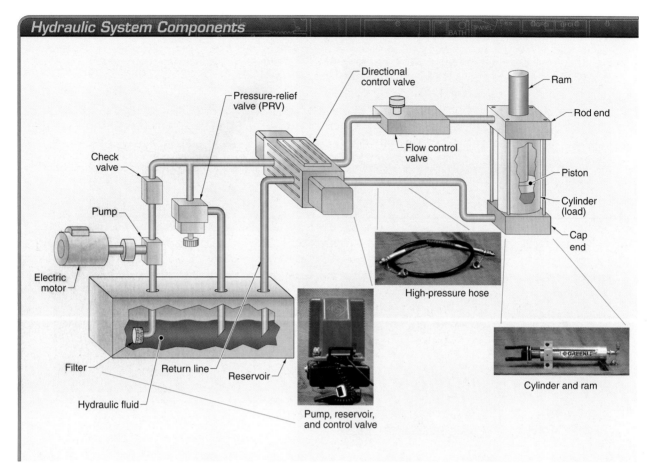

*Figure 6-1. Hydraulic systems all have common components. The specific design, however, may vary from bender to bender.*

## Bender Components

In addition to the hydraulic system, hydraulic benders have several components in common. **See Figure 6-2.** All hydraulic benders have a yoke to hold the cylinder and ram in the proper position, a pipe collar or strap to hold the conduit on the shoe, a shoe to form the bend, and a follow bar or rollers.

## One-Shot Benders

Two general categories of hydraulic benders are one-shot benders and multiple-shot benders. One-shot benders can fabricate a 90° bend in one operation. **See Figure 6-3.** The maximum bend of a one-shot bender is just over 90° to allow for springback. Modern hydraulic benders are usually one-shot models since they are more efficient and easier to use.

## Multiple-Shot Benders

Multiple-shot benders require two or more operations to fabricate a 90° bend. **See Figure 6-4.** The maximum bend of a multiple-shot bender is less than 90°; therefore, several bends are required to complete one larger bend. While one-shot benders are the most common, many multiple-shot benders do exist. In addition, one-shot benders are not available for 5″ or 6″ galvanized rigid conduit. Multiple-shot benders are required because of the physical size of the conduit and the large bend radius.

*Tech Fact*

Care should be taken when assembling a hydraulic bender because improperly connected parts can easily be bent or broken by the high pressure created by the pump.

**Hydraulic Bender Components**

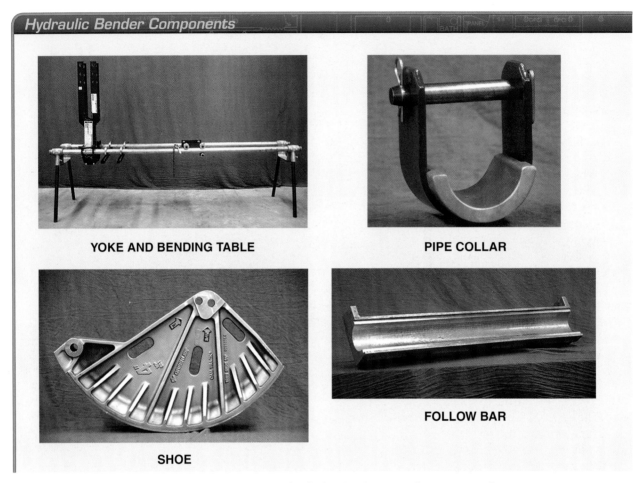

YOKE AND BENDING TABLE

PIPE COLLAR

SHOE

FOLLOW BAR

*Figure 6-2.* While varying greatly in type and style, all hydraulic benders have certain components in common.

**One-Shot Hydraulic Benders**

**Multiple-Shot Hydraulic Benders**

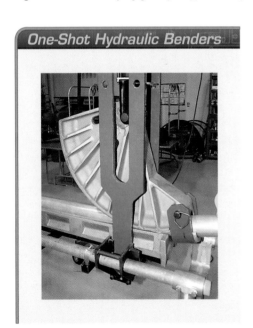

*Figure 6-3.* One-shot benders can fabricate a 90° bend in one operation.

*Figure 6-4.* Multiple-shot benders require two or more operations to fabricate a 90° bend.

## LAYOUT AND FABRICATION

All hydraulic benders must be set up before use. The equipment manual should provide important operating and safety instructions. A hydraulic bender should be set up in a dry area close to the work. It should be set up on a flat, level surface. The surface can be the floor or a table, but it must be flat and sturdy enough to support the bender and the weight of the conduit.

A hydraulic bender draws a large amount of current, so an adequate source of power must be located nearby. Extension cords and circuits must be properly sized to handle the load drawn by the bender.

### Layout

Layout for conduit bent with hydraulic benders is generally the same as for conduit bent with hand benders. However, placing pencil marks on large conduit can be more difficult than placing them on smaller conduit. The larger diameter makes it difficult to accurately draw the mark completely around the conduit. A strap or pipefitter's wrap-around can be used as a template to help draw the mark around the conduit when laying out bends. **See Figure 6-5.**

In the absence of a wrap-around, a steel measuring tape turned upside down and wrapped around the conduit can also be used to draw the layout lines. Also, there are many types of protractors on the market designed to aid the fabrication of bends. **See Figure 6-6.**

### Ram Travel

The key to accurate repeatable bends is ram travel. An electrician can make consistent, accurate bends in all types of field conditions with readily available tools by measuring the distance the ram travels out of the cylinder.

For any given shoe and type of conduit, any bend can be repeated by extending the ram the same distance for both bends. The amount of ram travel for a given bend angle can be measured, recorded, and used for future bends.

**Placing Pencil Marks**

*Figure 6-5. A strap or pipefitter's wrap-around can be used as a template to help mark the conduit when laying out bends.*

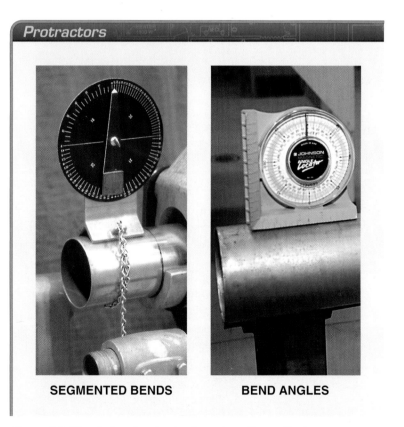

**Protractors**

**SEGMENTED BENDS**          **BEND ANGLES**

*Figure 6-6. The choice of protractor depends on the application.*

Some manufacturers have a chart located on the bender yoke that notes the ram travel for common bend angles for each shoe. **See Figure 6-7.** Other manufacturers have a rod that is attached to the ram that can be used to measure ram travel.

Some benders do not have any devices to assist in measuring ram travel. If this is the case, the ram travel can be found by using a tape measure or rule to measure the amount the ram has traveled out of the cylinder. **See Figure 6-8.** Once this distance is known, bends can be precisely repeated using this measurement. The ram travel should be measured after the conduit is bent but before the pressure is released.

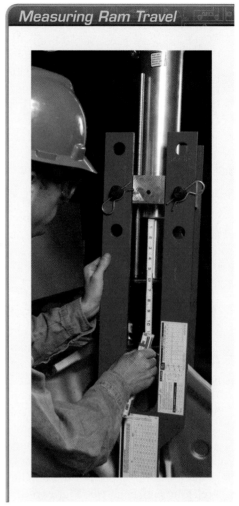

*Measuring Ram Travel*

**Figure 6-8.** *Ram travel can be found by using a tape measure or rule to measure the amount the ram has traveled out of the cylinder.*

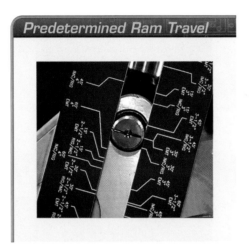

*Predetermined Ram Travel*

**Figure 6-7.** *Some manufacturers have a chart located on the bender yoke that notes the ram travel for common bend angles for each shoe.*

### 90° Bends

One of the simplest bends to make with a one-shot bender is a 90° bend. **See Figure 6-9.** Somewhere on the bender (or the bender box) there should be a chart noting the take-up for the size and type of conduit for each shoe. A 90° bend is fabricated as follows:

1. Measure the stub distance. Subtract the take-up and place a pencil mark on the conduit.

2. Place the conduit in the bender and align the pencil mark with the front of the shoe or pipe strap. Secure the conduit according to the manufacturer's instructions.

3. Start the pump and set the valve to allow the pressurized fluid to enter the cylinder. Bend the conduit until the bend approaches 90°.

4. Measure the amount the ram has travelled. Alternatively, place one level on the leg and a second level on the stub end of the conduit. Continue bending until the difference between the levels is 90°. Bend slightly past 90° to allow for springback.

EMT and rigid need only a small amount of overbend to allow for springback. With IMC, springback can be as much as 5° or 6° and can vary from piece to piece within the same bundle. Some trial and error may be required to get a true 90° bend.

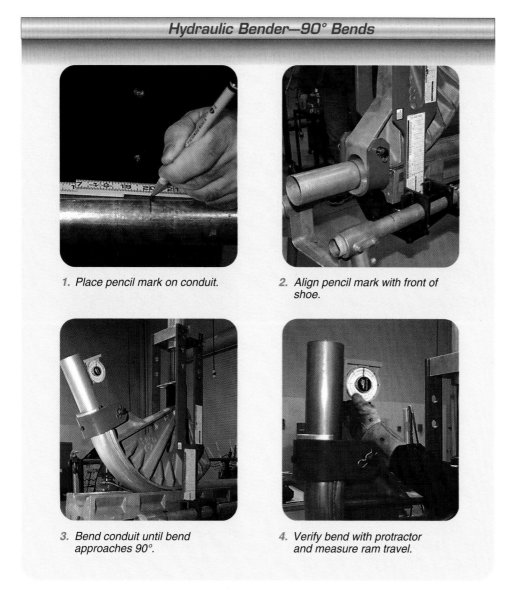

### Hydraulic Bender—90° Bends

1. Place pencil mark on conduit.

2. Align pencil mark with front of shoe.

3. Bend conduit until bend approaches 90°.

4. Verify bend with protractor and measure ram travel.

**Figure 6-9.** One of the simplest bends to make is a 90° bend with a one-shot bender.

## Offsets and Kicks

When bending offsets and kicks, both the multiplier method and the measured rise method can be used effectively. If the bender provided can accurately produce bends with a known angle, the multiplier method is very effective. **See Figure 6-10.** The procedure to fabricate an offset is as follows:

1. Measure the rise and calculate the shrink and distance between bends. Place pencil marks on the conduit at the proper locations.

2. Place the conduit in the bender and align the first pencil mark with the front of the shoe. Snug up the bender and place a no-dog device on the end of the conduit. Bend the conduit to the desired angle, measuring the ram travel to verify the angle.

3. Push the conduit through the bender, rotate the conduit 180°, and align the second pencil mark with the front of the shoe. Check the no-dog device to verify the conduit has rotated exactly 180°, and then snug up the bender.

4. Bend the conduit to the desired angle to complete the offset, measuring the ram travel to verify the angle.

If the bender provided cannot accurately produce a known angle, the measured rise method is the only practical way of fabricating accurate repeatable offsets. Ram travel can be measured to ensure the first and the second bends of an offset are bent to the same angle, even though the angle may be unknown.

A difficulty with the measured rise method of bending offsets occurs when running a rack with more than one size of conduit.

Ram travel is a good indicator of angle for a particular conduit type and size only. For different sizes of conduit, the angle cannot be duplicated by measuring ram travel.

There are several methods used for duplicating an angle on conduit of different sizes. The original angle may be duplicated using a protractor level, a folding rule, a piece of smaller EMT, or anything in which the desired angle can be reproduced. The different-sized conduit can then be bent to this angle, and a ram travel measurement can be taken. This measurement can then be used to reproduce the angle in the second offset bend.

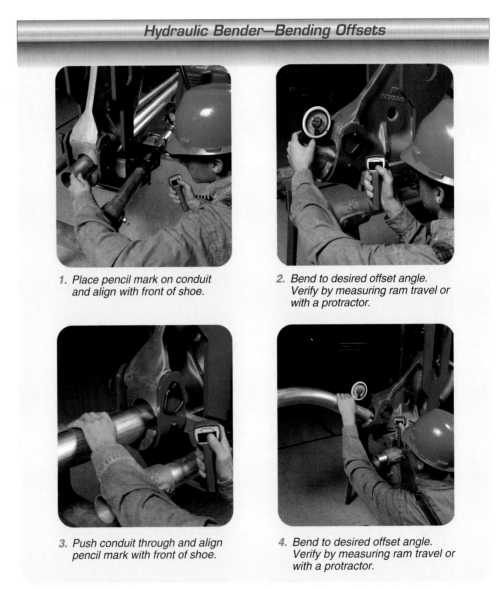

**Hydraulic Bender—Bending Offsets**

1. Place pencil mark on conduit and align with front of shoe.

2. Bend to desired offset angle. Verify by measuring ram travel or with a protractor.

3. Push conduit through and align pencil mark with front of shoe.

4. Bend to desired offset angle. Verify by measuring ram travel or with a protractor.

**Figure 6-10.** The multiplier method can be used to fabricate offsets with a hydraulic bender.

Kicks can also be produced using either the measured rise method or the multiplier method. **See Figure 6-11.** After the 90° bend of a kick is fabricated to the desired length, a kick can be fabricated with the measured rise method as follows:

1. Use a level to verify that the 90° stub is parallel to the ground.
2. Measure the height of the stub above the floor.
3. Bend the kick, using a tape measure or rule to measure the change in height above the floor.

If using the multiplier method, the kick is bent until the ram travel indicates that the desired degree of bend has been obtained. When bending larger sizes of conduit, the measurements should be made from the center of the stub instead of the back.

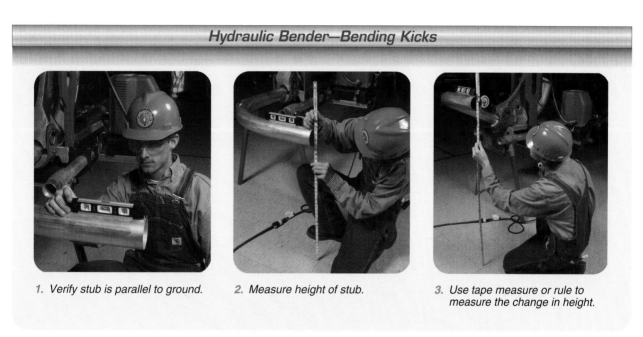

### Hydraulic Bender—Bending Kicks

1. Verify stub is parallel to ground.

2. Measure height of stub.

3. Use tape measure or rule to measure the change in height.

**Figure 6-11.** *The multiplier method can be used to fabricate kicks with a hydraulic bender. The height of the kick can be checked with a tape measure or rule.*

## SUMMARY

- When hydraulic pressure is applied to a piston in a pressurized vessel, force is transferred to the piston; if the piston is attached to a tool such as the ram of a bender, useful work can be done.

- The pressure and force developed in hydraulic benders can be very dangerous. A hydraulic bender should not be used unless the operation and risks are clearly understood.

- Hydraulic systems consist of pumps, valves, hoses, a cylinder and ram, and a fluid reservoir.

- Hydraulic benders consist of a yoke, a pipe collar or strap, a shoe, and a follow bar.

- A one-shot bender can fabricate a 90° bend in one operation.

- A multiple-shot bender requires two or more operations to fabricate a 90° bend.

- The key to accurate repeatable bends is ram travel. For any given shoe and type of conduit, any bend can be repeated by extending the ram the same distance for each bend.

# Other Conduit Types

## CONDUIT BENDING and FABRICATION

**7**

METALLIC CONDUIT ...................................................................................116
PVC-COATED CONDUIT ...........................................................................117
PVC CONDUIT...........................................................................................120
SUMMARY .................................................................................................123

Although the great majority of conduit used today is steel, there are other types of approved and available metallic and nonmetallic conduit. Steel rigid conduit is widely available and recognized as acceptable for raceway systems in most applications. However, rigid conduit may also be manufactured from stainless steel, brass, and silicon-bronze. Aluminum conduit is allowed in some applications.

PVC-coated steel conduit is used in corrosive environments. In order to protect the PVC coating, special tools are required. Standard clamping, cutting, and threading tools can damage the PVC coating. Modified tools are available to minimize the damage.

Nonmetallic PVC conduit has increased in use since its introduction in the early 1960s. PVC conduit is often used for underground installations and grounding electrode conductor raceways. Because PVC conduit is made of plastic, it has several advantages over metallic raceways. Voltage loss due to electromagnetic induction is nonexistent. PVC is impervious to the elements and does not easily corrode. It is much lighter in weight than rigid and is much easier to cut. Sections of PVC conduit are solvent-welded together and do not require threading.

## OBJECTIVES

1. List common types of metallic conduit.
2. Describe the differences between standard clamping, cutting and threading, and bending tools, and tools used with PVC-coated conduit.
3. Describe how sections of PVC conduit are joined together.
4. Describe how to bend PVC conduit.
5. List the different types of PVC heaters.
6. Explain why it is necessary to rotate PVC conduit while it is in a heater.

## METALLIC CONDUIT

In addition to the conventional types of conduit, there are several other types of metallic conduit available. The most common of these is aluminum conduit. Metallic conduit can also be made from stainless steel, silicon-bronze, and brass. Several types of PVC-coated steel conduit are also available.

### Aluminum Conduit

Aluminum conduit is extremely lightweight compared to steel conduit. **See Figure 7-1.** It provides an excellent ground path and it can be bent with virtually no springback. The disadvantages of aluminum conduit include less physical protection for conductors because aluminum is softer than steel.

Aluminum conduit is typically used where weight is a concern, such as when the conduit needs to be installed above a finished ceiling. In addition, aluminum conduit has much lower impedance than steel conduit and therefore makes a better equipment grounding conductor.

> **Tech Tip**
>
> Thread/joint lubricants may be used with aluminum conduit to prevent seizing and galling and to reduce wear and breakage of mating parts.

| | Conduit Weight | | |
|---|---|---|---|
| Size | Galvanized Rigid Conduit* | EMT* | Aluminum* |
| ½" | 82 | 30 | 28 |
| 1" | 161 | 67 | 55 |
| 2" | 350 | 148 | 119 |
| 4" | 1030 | 393 | 350 |

*weight in lb. per 100'

**Figure 7-1.** Aluminum is the lightest metallic conduit.

Aluminum is easy to cross-thread when installing a coupling or fitting. Also, aluminum conduit does not shield wiring from outside electromagnetic interference (EMI) as well as steel conduit. In industrial installations where critical digital and analog control conductors require extensive shielding from high-power noise, such as from motors or heaters, the conduit is normally made of steel or other ferrous (iron-based) metal.

Aluminum conduit is available with wall thicknesses corresponding to rigid and EMT steel conduit. Threaded aluminum conduit is readily available in 10' lengths, including the coupling, in trade sizes from ½" to 6". Plastic thread protector caps are color coded to help with identification. The cap colors are the same as for conventional rigid conduit—blue for even inch sizes, black for ½" sizes, and red for ¼" sizes. **See Figure 7-2.**

| Aluminum Thread Protector Caps | | |
|---|---|---|
| Color | Sizes | Examples |
| Blue | Inch sizes | 1", 2", 3", 4", 5", 6" |
| Black | ½" sizes | ½", 1½", 2½", 3½" |
| Red | ¼" sizes | ¾", 1¼" |

**Figure 7-2.** Plastic thread protector caps are color coded to help with identification.

### Other Metallic Conduit

Rigid conduit can be manufactured from a variety of metals other than steel. Depending on the requirements, the designer may specify stainless steel, silicon-bronze, or brass conduit. Stainless steel conduit is corrosion resistant, but extremely hard and very difficult to bend, cut, and thread. Also, stainless steel conduit has quite a bit of springback. In addition, stainless steel conduit must be threaded slowly with plenty of cutting oil.

Silicon-bronze alloys may also be chosen for improved corrosion resistance. Brass conduit, while considerably softer than stainless, also has quite a bit of springback. With brass conduit, it is not unusual to have to bend 105° to end up with a 90° bend.

> **Tech Fact**
>
> Aluminum conduit is often used when the wiring in the conduit is carrying high-frequency (400 Hz) alternating current, such as in military, data processing, and radar installations, where steel raceways would be subject to inductive heating.

## PVC-COATED CONDUIT

PVC-coated conduit is often specified for use in corrosive industrial environments. PVC-coated conduit is usually installed as a system where all of the raceway components (couplings, fittings, enclosures, straps, and hangers) are coated with PVC. PVC-coated conduit is commonly called robroy, after Robroy Industries, a manufacturer of PVC-coated conduit.

### Conduit Sizes

Most often, the PVC coating is applied to standard rigid conduit, although the coating can also be applied to IMC and EMT. Threaded PVC-coated rigid conduit is readily available in 10′ lengths, including the coupling, in trade sizes from ½″ to 6″. Plastic thread protector caps are color coded to help with identification. The cap colors are the same as for conventional rigid conduit—blue for even inch sizes, black for ½″ sizes, and red for ¼″ sizes.

### PVC Coating

The PVC coating on the outside of the conduit is 40 mils (0.040″) thick. This coating increases the outside diameter of the conduit by 80 mils. The PVC coating is applied over three types of conduit. These three types of conduit are bare steel, galvanized steel, or zinc-coated steel. **See Figure 7-3.**

The outer coating comes in a variety of colors. The most common coating is gray. The conduit manufacturer's specifications should be consulted when selecting the correct coating for the installation. The inside of the conduit has a 2-mil urethane coating.

**Coating Repair.** PVC-coated conduit is used in corrosive environments. Cutting the threads exposes the steel conduit, and bare steel is vulnerable to corrosion. Therefore, after the threads are cut, they must be thoroughly degreased, cleaned, and dried. Any standard degreasing spray or liquid can be used. After the threads are cleaned, a touch-up compound must be applied to the threads. This restores the corrosion protection to the threads.

The PVC coating can become damaged despite attempts to keep it intact. Any areas with exposed metal should be repaired as soon as possible. Patching material may be supplied with the original conduit order. A corrosion inhibitor should be used to paint over any exposed areas. Several coats may be necessary. Coatings for this purpose are listed under UL (Underwriters Laboratories®) category "FOIZ".

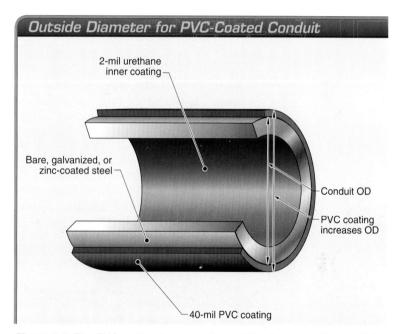

**Figure 7-3.** The PVC coating increases the outside diameter of the conduit.

### Tools

Bending PVC-coated conduit requires different tools and techniques than when bending rigid conduit. The main difference between PVC-coated conduit and other types of metallic conduit is that the PVC coating must be protected from damage. This means that the bending and handling tools need to be modified or special tools need to be used that do not damage the coating. These tools include clamping tools, cutting and threading tools, and bending tools.

**Clamping Tools.** Clamping tools, such as pipe vises and wrenches, must not have teeth that can damage the PVC coating. A common modification is to replace the jaws of a pipe vise with a three-sided adapter that spreads the clamping force over a larger area. **See Figure 7-4.**

If a pipe vise cannot be used, a half-shell clamp can hold the coated conduit. This clamp also spreads the force over a larger area. A half-shell clamp can either be purchased or can be made from a section of rigid conduit.

To make a half-shell clamp, two sections of rigid conduit are used, each about 6″ long in the next larger trade size. The two sections of rigid conduit are cut lengthwise with a band saw, with the cuts placed slightly off center. The larger halves are discarded, and the smaller halves are kept to use as the half-shells. The two halves should not touch each other when clamped around the PVC-coated conduit. A half-shell clamp can also be used with a chain vise.

**Cutting and Threading Tools.** PVC-coated conduit can be cut with any conventional cutting tools, such as a rolling cutter or a saw. As with any other type of conduit, the cut end must be reamed before use.

PVC-coated conduit can be threaded with standard threading tools. However, the outside diameter of PVC-coated conduit is larger than standard conduit of the same trade size because of the thickness of the PVC coating. Therefore, the PVC coating on the end must be pencil-cut (bevel cut) before threading. **See Figure 7-5.** This allows the die teeth to engage the conduit.

The PVC coating needs to be cut in a way that allows the threading die to remove the coating in small pieces during the threading operation. This will prevent long strips of the PVC coating from fouling the die. A thread protector cap is used to determine the length of the threads, and a cut is made around the circumference at that point. Longitudinal cuts in the coating are made from that point to the end of the conduit where the threads will be cut.

---

*T*ech Fact

Care should be taken when using degreasing solvents and solvent cements. Proper PPE is needed, and the MSDS requirements must be followed.

---

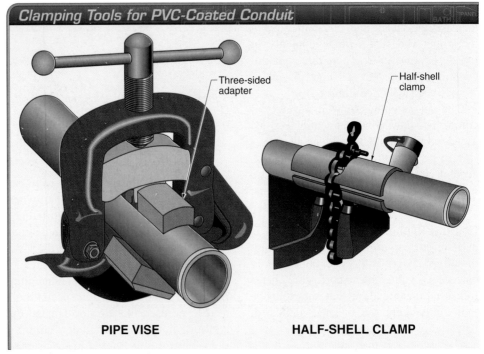

**Clamping Tools for PVC-Coated Conduit**

Three-sided adapter

Half-shell clamp

**PIPE VISE**    **HALF-SHELL CLAMP**

*Figure 7-4. PVC-coated conduit requires special clamping tools to prevent damage to the coating.*

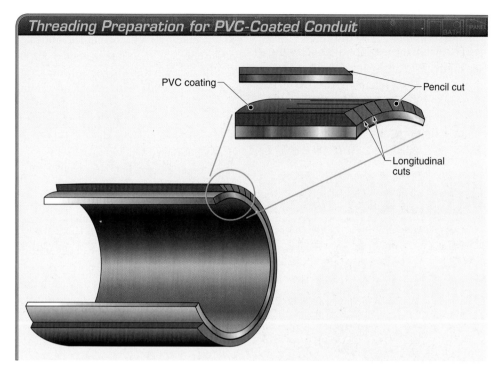

**Figure 7-5.** *The PVC coating must be prepared before threading.*

**Bending Tools.** Because of the coating, the outside diameter of PVC-coated conduit is slightly larger than standard conduit. Therefore, it cannot be bent with standard bending shoes without damaging the coating.

PVC-coated conduit must be bent on special shoes designed to bend this particular type of conduit. **See Figure 7-6.** For hand bending, some manufacturers recommend using the next-larger-size bender. Existing shoes should not be modified to accommodate the coating. Shoes are manufactured to fine tolerances and any field modification will ruin the shoe.

## Couplings and Fittings

Sleeves are present on couplings and fittings for PVC-coated conduit. These sleeves are used at all joints to protect the joint from corrosion. The installation of couplings and fittings can be a little difficult in cold weather. It is much easier to place the sleeve when it has been warmed before installation. An easy way to warm the sleeves is to immerse them in warm water until they are soft and pliable.

**Figure 7-6.** *Special bender shoes are required when bending PVC-coated conduit.*

Strap wrenches or pipe wrenches with specially equipped wide jaws should be used to install these couplings and fittings. **See Figure 7-7.** Slip-joint pliers or standard pipe wrenches should not be used to install PVC-coated conduit or fittings. Threadless fittings must not be used with PVC-coated rigid conduit or IMC.

### Tech Tip

Only fresh solvent cement should be used on joints. If solvent has started to gel or get stringy, a joint can fail and require costly repairs. The cement should be applied evenly over the outside of the conduit, and a thin coat should be applied on the inside of the fitting.

**Modified Tools for PVC-Coated Conduit**

*Figure 7-7. Standard tools must be specially equipped for handling PVC-coated couplings and fittings.*

## PVC CONDUIT

Nonmetallic PVC conduit is a very different raceway than the other raceways discussed so far. Because PVC conduit is made of plastic, it has any number of advantages over ferrous raceways. Voltage loss due to electromagnetic induction is nonexistent. PVC is impervious to the elements and does not easily corrode. It is lighter in weight than rigid, and much easier to cut. Sections of PVC conduit are solvent-welded together and do not require threading.

### PVC Conduit Sizes

PVC is available in a number of wall thicknesses. These are designated in schedules. Schedule 20 has a very thin wall and is used in underground installations when encased in concrete. Schedule 40 is the standard wall thickness and is used both aboveground and belowground. Schedule 80 has a very heavy wall and is often used in direct burial situations under heavy traffic areas such as roads, streets, and parking lots.

## Joining PVC Conduit

All schedules of PVC conduit are fabricated and installed by the same methods, despite the differences in wall thickness. Sections are connected together with a solvent-welding technique. The solvent is a powerful chemical and must be used in a well-ventilated area. The material safety data sheet (MSDS) and all precautionary labels must be read and understood before using the solvent.

There are several types of UL-listed solvent cements on the market for different temperature ranges and conduit sizes. **See Figure 7-8.** The labels must be read carefully. If a warm-weather solvent is used in cold weather, the solvent will not dry properly and may attack the conductor insulation. If a cold-weather solvent is used in warm weather, the cement will cure before the joint can be assembled.

**PVC Solvent**

*Figure 7-8. Sections of PVC conduit are connected with a solvent-welding technique.*

Mating surfaces should be dry and clean before applying the solvent. Both pieces being joined should be coated with the solvent. The solvent works by dissolving the PVC on the surface of the conduit and fittings where they will be placed in contact with each other. After the connection is made, the joint needs to be rotated a quarter turn to make sure all the mating surfaces are in complete contact.

Depending on the temperature, the joint may set up in just a few seconds or a minute. As the solvent dries, the two objects are fused (welded) together. This produces an extremely strong joint. Once the joint has cured, it cannot be taken apart.

PVC can be cut in any number of ways. As with metallic conduit, PVC conduit should be reamed to remove any burrs or rough edges. If the conduit is of a relatively small diameter, a ratchet cutter can be used to make quick, clean cuts. **See Figure 7-9.** For larger-diameter PVC conduit, handsaws are available that produce clean and easy cuts. Regardless of the cutting method, the joint must be cut square and any burrs must be removed.

PVC can also be cut with a good quality braided mason's string. This saves the electrician time when attempting to cut conduit in a trench. The mason's string is placed around the conduit and the string is pulled back and forth with both hands.

## Bending PVC Conduit

Bending PVC conduit is a somewhat challenging task. PVC conduit is very rigid and cannot be bent at room temperature. The conduit must be heated before it will bend. However, PVC retains heat for some time. This can cause serious injury if proper care is not taken.

After the conduit reaches bending temperature, it can be bent around a template. **See Figure 7-10.** A bending template can be as simple as a piece of EMT bent to specifications or as complex as a sheet of plywood with blocks secured to it to define the proper bend. Simple bends in PVC conduit can also be made by hand without a template. Care must be taken to ensure that the conduit does not kink or distort as it is being bent.

Once the conduit has been heated and formed into the proper bend, it must be held in the new shape until it cools. Water can be applied with rags or a sponge to speed up the cooling. This helps accelerate the entire bending procedure. If water is not used, it may take some time for the conduit to cool.

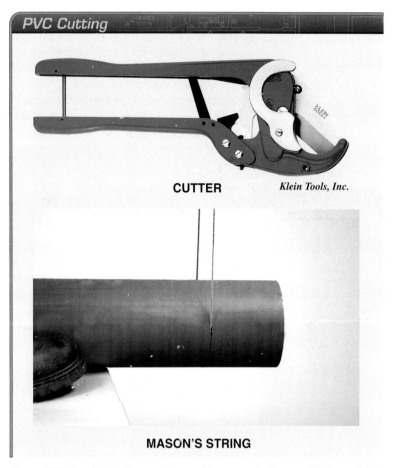

**PVC Cutting**

**CUTTER** *Klein Tools, Inc.*

**MASON'S STRING**

*Figure 7-9. A ratchet cutter can be used to cut small-diameter PVC conduit. A mason's string makes it easy to cut larger sizes of PVC conduit in the field.*

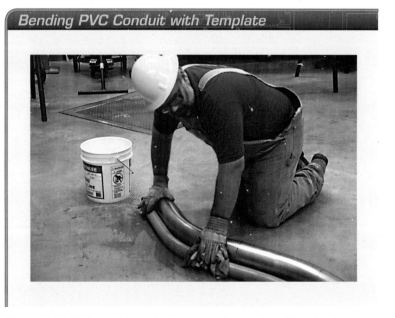

**Bending PVC Conduit with Template**

*Figure 7-10. PVC conduit can be bent around a template. Water is often used to cool the conduit after it is formed to the correct shape.*

Smaller-diameter PVC conduit can be easily bent to any angle using this process. Larger-diameter PVC conduit is more difficult to bend because an extremely large bend radius is required. Standard factory-made sweep elbows are available to replace a field bend. The use of elbows makes it unnecessary to bend the conduit and can speed up a job.

### PVC Heaters

There are a number of tools available to heat PVC. They all have advantages and disadvantages. The most common heaters are hot boxes. Heating blankets or pads that wrap around the conduit and plug into an outlet can also be used. Gas torches can be used with a special nozzle to create a wide, diffuse flame to evenly distribute the heat. Glycol heaters have also been used but are seldom employed anymore.

Pipe plugs may be placed in the ends of the conduit to help the heating process along. Pipe plugs keep the hot air inside the conduit rather than allowing it to escape out the ends. The air trapped inside the conduit also helps keep the conduit from kinking or collapsing while it is being bent. This helps meet the NEC® requirement that rigid nonmetallic conduit not be damaged, nor its diameter reduced when making bends. **See Figure 7-11.**

**PVC Pipe Plugs**

*Figure 7-11. Pipe plugs are used to keep heated air within the conduit and speed up the bending process.*

**Tech Tip**

PVC conduit should be checked for a dry fit before beginning. A loose fit may not cement together correctly. The conduit should not bottom out in the socket.

**Hot Boxes.** A *hot box* is a PVC heating tool containing an electric heating element within an enclosure that holds the heat. **See Figure 7-12.** Hot boxes range from relatively small units designed to bend ½″ conduit to larger units for bending 4″ conduit. A small hot box may use 110 V, while a large hot box may need 220 V.

The key to using a hot box is to keep the conduit turning in the enclosure. If the pipe is not rotated, the side toward the heating element will overheat and blister. Rotation keeps the heat uniformly distributed over the entire bending area of the conduit. Some models have an electric roller system that turns the conduit automatically.

Hot boxes are reliable, easy-to-use machines. Their main disadvantage is that they require a power supply. Electric power may not be readily available when installing a large underground conduit system far from the electric utility.

**Torches.** If electric power is not available, there are a number of torches available for heating PVC. These torches need to have an extremely wide flame to evenly distribute the heat and must be identified as suitable for the purpose. The key to using these torches to heat PVC is to keep the flame moving over the entire area to be bent. This ensures that the entire section will be heated properly and the conduit will not blister.

**Glycol Heaters.** Glycol heaters have not been available from manufacturers for some time but are still occasionally used in the field. These heaters used an electric or LP gas heating unit to heat glycol and pump it through a blanket placed around the conduit. Glycol holds heat extremely well, and if the glycol escapes the system, it can cause severe burns. If a glycol heater is being used on a job site, the electrician must understand the operational procedure before using it.

**Tech Fact**

Work quickly when joining PVC joints together. Assemble the joints immediately and hold them in place for about a minute until they begin to set. Wipe off any excess cement.

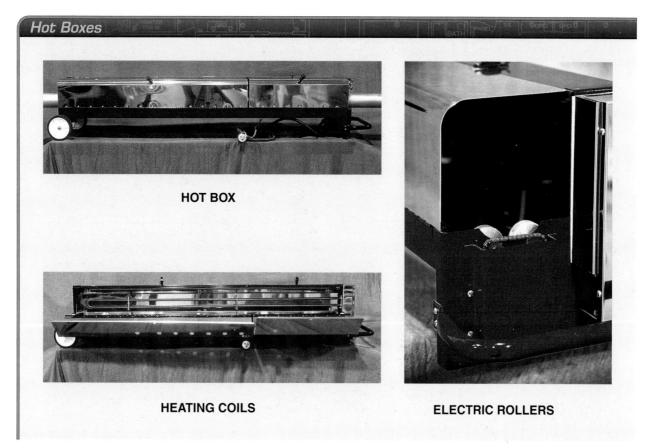

**Hot Boxes**

**HOT BOX**

**HEATING COILS**

**ELECTRIC ROLLERS**

*Figure 7-12.* A hot box can be used to heat PVC conduit for bending. A heating coil provides the heat, and an electric roller rotates the conduit for even heating.

## SUMMARY

- In addition to the conventional types of conduit, metallic conduit is available in aluminum, stainless steel, silicon-bronze, and brass.

- PVC-coated conduit is often specified for use in corrosive industrial environments.

- For PVC-coated conduit, a common tool modification is to replace the jaws of a pipe vise with a three-sided adapter that spreads the clamping force over a larger area.

- If the PVC coating becomes damaged, it must be repaired with touch-up compound or special paint.

- Sections of PVC conduit are joined together.

- PVC conduit must be heated before bending.

- A hot box contains an electric heating element within an enclosure.

# Threaded Conduit

## CONDUIT BENDING and FABRICATION

**8**

THREADERS .................................................................. 126
THREADED CONDUIT ....................................................... 133
SUMMARY ................................................................... 137

**M**any types of conduit need to be threaded. The tools available to thread conduit are varied in complexity and ease of use. These tools range from simple hand-driven threaders to expensive large power threaders. Each type of threader is used with a particular set of field conditions.

A die is a cutting tool used to form external screw threads on conduit. Dies are precision tool pieces made of hardened steel and must be held in exact alignment with conduit for threads to be properly cut. Handheld threaders use a die head that rotates on the conduit. Threading machines rotate the conduit while the die head is fixed.

Thread length and taper are important factors when connecting couplings and fittings to conduit. Conduit is threaded with tapered NPT threads. Couplings and fittings are threaded with straight NPT threads. Threads must be cut to the correct length to prevent damage and poorly fitting conduit.

Threading conduit can be very difficult if the conduit has already been bent. The bends in the conduit may not allow it to be chucked into a chain vise. Determining gain and shrink can allow an electrician to correctly cut conduit to length and thread both ends of the conduit.

## OBJECTIVES

1. List the components of a conduit threading tool.
2. Explain the difference between a ratcheting and a nonratcheting threader.
3. Describe the necessary safety precautions when using a power-driven threader.
4. Explain why threads must be cut to the correct length.
5. Demonstrate how to thread conduit with a handheld threader and with a threading machine.

## THREADERS

A *conduit threader* is a tool used to cut threads in conduit. Conduit threaders vary in both complexity and ease of use. They range from simple hand ratchets to expensive power-driven equipment. **See Figure 8-1.**

Hand tools are simple to operate, but they require time and a lot of physical energy. Power tools are more complex in their operation and need electricity, but they save time and require only a small amount of physical energy.

---

**Tech Fact**

Used cutting oil is an environmental hazard and must be disposed of properly.

---

**Threading Tools**

**HAND RATCHET**

RIDGID

**POWER-DRIVEN EQUIPMENT**

*Ridge Tool Company*

**Figure 8-1.** *The tools available to thread conduit are varied in both complexity and ease of use. They range from simple hand ratchets to expensive power-driven equipment.*

## Conduit Threader Components

There are several components common to all conduit threaders. The components are dies and die heads, oilers, and reamers.

**Dies and Die Heads.** A *die* is a cutting tool used to form external screw threads in conduit. Dies are precision tool pieces made of hardened steel. All threaders use at least four die pieces. **See Figure 8-2.**

Dies must be held in exact alignment with each other and with conduit for threads to be cut properly. A *die head* is the part of a conduit threader that holds the dies securely in position and applies pressure to cut external threads. A die head is also called a diestock. Die heads may be attached to a handle and rotate on fixed conduit, or they may be attached to a threading machine with conduit that rotates. Since dies eventually wear or break, die heads are designed to be easy to disassemble. The end cap of a die head can be removed to replace the dies.

A *quick-opening die head* is a die head with a release lever at the top that can be raised manually to retract the dies and release the conduit after the thread has been cut. Quick-opening die heads are available as mono or adjustable die heads. These die heads are often used on threading machines. A mono die head is used to thread only one size of conduit and requires different dies for each size of conduit. An adjustable die head is used to thread several conduit sizes with the same dies.

**Oilers.** An *oiler* is a device used to apply cutting oil to the dies during the thread cutting process. Most oilers also separate metal shavings from the used oil and capture the oil for reuse. **See Figure 8-3.** Specifically formulated cutting oils should be liberally applied during the thread cutting process to keep die wear to a minimum. If dies are used without cutting oil, they will be ruined.

Oil is applied using an oiler, oil can, or oiler attachment mounted on an oil container. Simple hand oilers are often used to apply cutting oil. In some cases, threaders have automatic oilers.

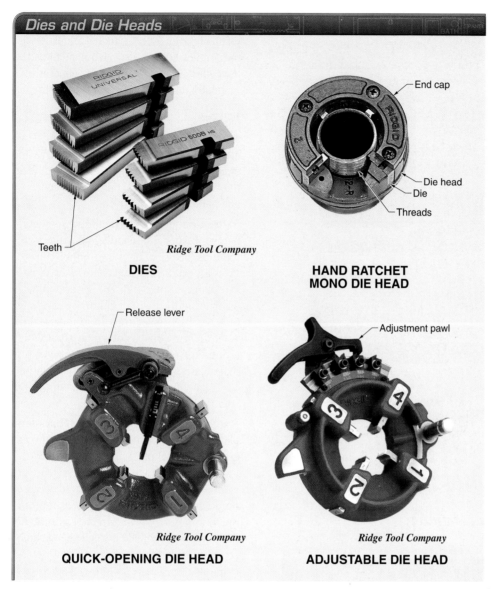

**Dies and Die Heads**

*Ridge Tool Company*

Teeth

**DIES**

End cap
Die head
Die
Threads

**HAND RATCHET MONO DIE HEAD**

Release lever

*Ridge Tool Company*

**QUICK-OPENING DIE HEAD**

Adjustment pawl

*Ridge Tool Company*

**ADJUSTABLE DIE HEAD**

*Figure 8-2. Dies are held in a die head. Die heads come in a variety of configurations.*

**Reamers.** When conduit is cut, rough and sharp edges that can damage wire insulation often remain. A *reamer* is a tool used to remove burrs and sharp edges from a piece of conduit after it has been cut to length. **See Figure 8-4.** A reamer consists of a handle, grip, and head that remove the sharp edges on the conduit. Reamers may be small handheld tools or they may be mounted on a carriage on a larger threading machine.

Improper reaming can ruin threads. The conduit should be reamed before being threaded. The threads can be deformed if the conduit is reamed after the threads are cut. The sharp edges and burrs should be removed, but the conduit should not be excessively reamed. Excessive reaming causes the conduit end to become deformed into a bell shape, after which the dies will not seat properly.

**Tech Fact**

Pipe cutters are used to cut conduit. A pipe cutter contains a sharp cutting wheel that is slowly turned while being pressed against the conduit.

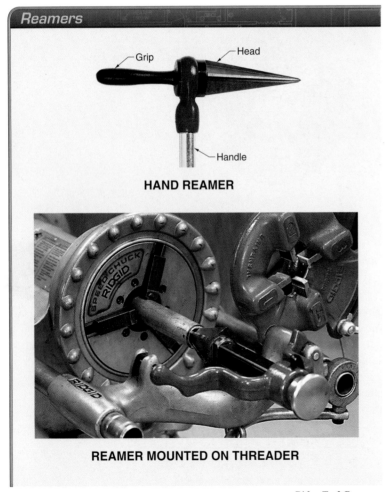

**Oilers**

Manual oiler

Reservoir

**MANUAL OILER**

**CUTTING OIL**

*Ridge Tool Company*

**Figure 8-3.** *An oiler is a device used to apply oil to dies and capture used oil in a pan or reservoir for recirculation.*

**Reamers**

Grip

Head

Handle

**HAND REAMER**

**REAMER MOUNTED ON THREADER**

*Ridge Tool Company*

**Figure 8-4.** *Reamers may be handheld or they may be mounted on a threader.*

## Handheld Threaders

Of all of the threaders, handheld threaders are the simplest tools to use. Handheld threaders use a die that is rotated on the conduit by a manually turned handle or by an electrically powered drive. They may be used when the electrician does not have the time and space to set up power threading equipment. They are fairly mobile and do not involve the cumbersome setup of power tools. Handheld threaders may be ratcheting or nonratcheting, and hand- or power-driven.

**Vises.** Conduit must be held securely while being threaded. For handheld threaders, this requires the use of a vise. Common types of vises are yoke vises and chain vises. A *yoke vise* consists of a tightening screw that is used to tighten a yoke holding a set of jaws that are used to securely hold the conduit. A *chain vise* consists of a chain that wraps around the conduit and a tightening screw that secures the chain around the conduit. Vises may be mounted on a bench or on a stand. **See Figure 8-5.**

**Ratcheting and Nonratcheting Threaders.** Conduit threaders may be ratcheting threaders or nonratcheting threaders. A *ratchet threader* is a conduit threader with a ratchet mechanism built into the handle and requires only a small amount of clearance for proper operation. A *drophead threader* is a ratchet threader in which the entire die head can be replaced with another die head for different conduit sizes. The reversible ratchet knob is pulled and the die head is dropped out of the threader. **See Figure 8-6.**

A *nonratcheting threader* is a conduit threader in which the handles are rotated completely around the conduit to turn the die head. This requires adequate clearance for the handles. A *three-way threader* is a nonratcheting threader that holds three die sizes simultaneously. A three-way threader typically consists of a ½″, ¾″, and 1″ die size combination.

**Tech Fact**

Replacement dies must be installed in a particular order. Individual dies are numbered and must be matched with their corresponding slots in the die head.

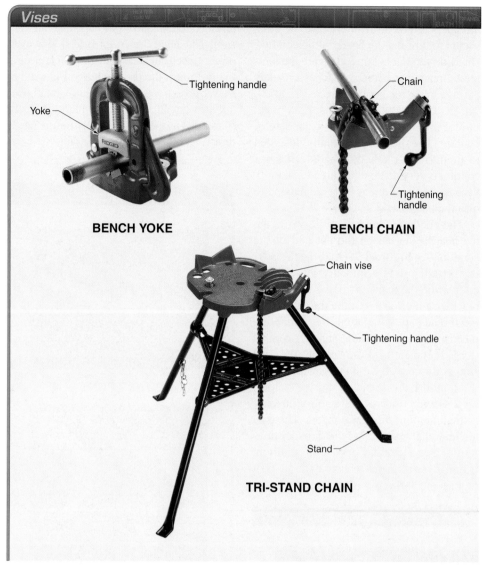

*Figure 8-5. Vises are used to hold conduit while it is being threaded.*

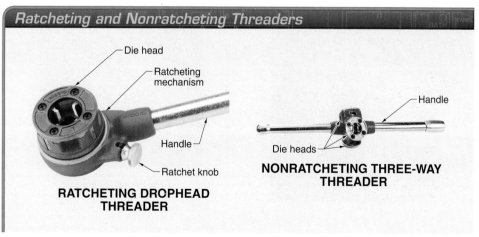

*Figure 8-6. Threaders come in ratcheting and nonratcheting designs.*

**Hand-Driven Threaders.** Hand-driven threaders use the arm strength of the electrician to turn the die head on the conduit. Hand-driven ratchet threaders are the simplest threaders to operate. Ratchet threaders are available for conduit from ½″ to 2″. They consist of a die head that holds a set of dies arranged for one particular size of conduit. The die head is usually designed to fit into the ratcheting handle. Larger conduit is very difficult to thread manually because of the amount of force needed to turn the threader.

The conduit is held in a vise while the die head is started on the cut end of the conduit. **See Figure 8-7.** Once the threader is started, oil is applied from a hand oiler while the handle is being turned in a clockwise motion to cut the threads. Once the threads are completed, the ratcheting mechanism on the handle can be reversed (turned counterclockwise), and the die head simply unscrewed from the conduit.

Hand-driven threaders can be helpful on a service truck or on a job with only a small amount of rigid conduit. The dies are interchangeable, and the threader does not require electrical power. In addition, hand-driven threaders are small and do not require much room for storage.

**Power-Driven Threaders.** Power-driven threaders, often called "power ponies," are very similar to hand-driven threaders. However, power-driven threaders use a reversible electric motor to turn the die head. **See Figure 8-8.** Power ponies require the use of a hand oiler and can thread ½″ to 2″ conduit. They are often used as the power supply for die heads directly mounted on large conduit.

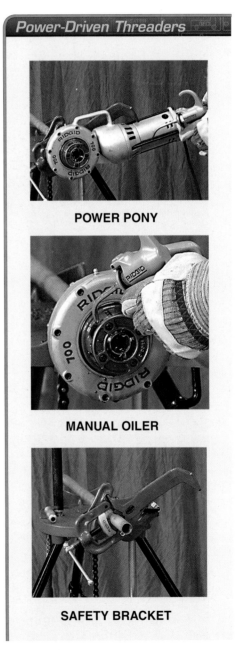

**Power-Driven Threaders**

**POWER PONY**

**MANUAL OILER**

**SAFETY BRACKET**

*Figure 8-8.* A power pony needs a manual oiler and should be used with a safety bracket.

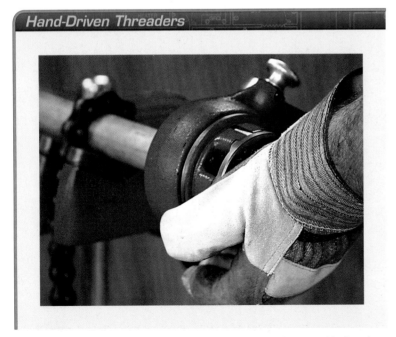

**Hand-Driven Threaders**

*Figure 8-7. Conduit is held in a vise while the die head is started by hand.*

Power-driven threaders are high-torque tools and can cause serious injury if not used correctly. Many manufacturers provide a safety bracket that mounts onto conduit to prevent the threader from moving. In addition, some threaders have a boss or hub in the end of the handle where a short piece of 1″ rigid conduit can be inserted or screwed for additional leverage to help control the tool.

Power-driven threaders are often used on jobs where a lot of threading is done but where there may not be space available to set up a fabrication shop. They are also portable enough to be used on projects that require the electrician to move many times throughout the workday.

### Large Power Threaders

Large power threaders are different from hand- and power-driven threaders. Instead of the rotating die head found on handheld threaders, power-driven threaders have a fixed die head and rotate the conduit. Large power threaders require a stand or bench to hold the tool. They can be very dangerous due to the amount of torque they produce. The most common type of large power threader is the threading machine.

**Threading Machines.** Threading machines are available to thread conduit from ½″ to 4″. Threading machines range from simple threaders that have manual oilers but do not have cut-length-limiting devices, to complex threaders with automatic oilers and devices to cut standard thread. The simplest setup uses a threading machine with a hand threader. Die heads on threading machines often are adjustable for varying sizes of conduit. Typically, one die head can be adjusted to cut threads on a particular range of conduit sizes. **See Figure 8-9.**

A threading machine has chucks located at the rear and front of the threader. When conduit is placed in the threader, the chuck at the rear is closed first. The chuck at the front is closed last. A threading machine setup usually includes a conduit cutter that is attached to the threading machine stand. Conduit that is fastened to the front chuck of the threading machine can be cut to size by using the conduit cutter. **See Figure 8-10.**

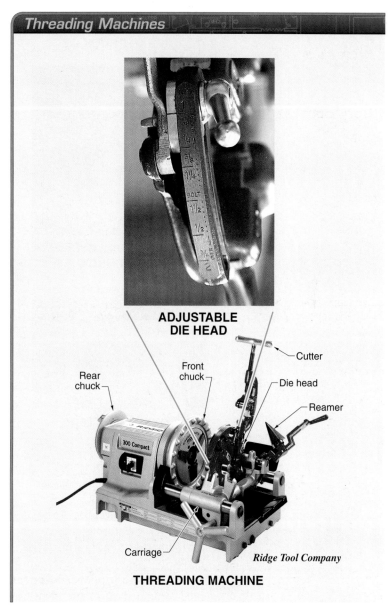

**Threading Machines**

ADJUSTABLE DIE HEAD

Rear chuck · Front chuck · Cutter · Die head · Reamer · Carriage

*Ridge Tool Company*

THREADING MACHINE

*Figure 8-9. A threading machine is often used for threading conduit. Many include adjustable die heads to simplify their setup.*

## Threading Machine Components

**REAR CHUCK**

**FRONT CHUCK**

**CONDUIT CUTTER**

*Figure 8-10. Threading machines include rear and front chucks to hold the conduit while being threaded and a cutter to cut the conduit to length.*

Conduit is typically cut and threaded on the job site with a threading machine. A small-diameter threading machine is designed for 2″ and smaller conduit, and is mounted on a tripod stand. It can be moved around the job site easily. A large-diameter threading machine is designed for 4″ conduit and smaller, and is typically used in a prefabrication shop or on a job where the machine can be moved between floors on a lift or elevator. Large-diameter machines typically have an integral oil reservoir and oil pump to lubricate the threads or grooves while they are being cut.

**Accessories.** There are accessories available for threading machines that make some parts of the job easier to accomplish. A conduit support is an accessory used to support long lengths of conduit and must be used with a threading machine to prevent the lengths of conduit from tipping the machine over. **See Figure 8-11.** A very convenient accessory is a close nipple chuck, which allows for a nipple (short piece of conduit) to be fabricated. The close nipple chuck can save a lot of time on a project by helping to fabricate nipples.

## Threader Accessories

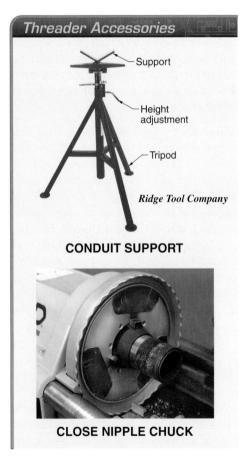

Support

Height adjustment

Tripod

*Ridge Tool Company*

**CONDUIT SUPPORT**

**CLOSE NIPPLE CHUCK**

*Figure 8-11. Common threader accessories include conduit supports and close nipple chucks.*

# THREADED CONDUIT

In order to attach multiple sections of conduit together, some types of conduit must be threaded. Once a section of conduit is threaded and bent (if necessary), the male threads of the conduit must be able to fit into the female threads of one end of a coupling or fitting in order to fasten the conduit and fitting together. Then, another section of conduit is fastened to the opposite end of the same fitting to combine two sections of conduit.

## Thread Length and Taper

Thread length and taper are important factors when installing threaded conduit fittings. In order for conduit to fit properly into a coupling or fitting, a standard thread must be cut into the conduit. National Pipe Taper (NPT) threads are used on conduit and fittings. The *National Pipe Taper (NPT) thread* is a standard thread used for connecting conduit in which the adjoining sides of the threads are at a 60° angle to each other.

Smaller conduit sizes typically need more threads per inch than larger conduit sizes. Different conduit sizes that need the same number of threads per inch can usually be threaded with the same dies using an adjustable die head.

It may prove worthwhile to check the die head on an unfamiliar machine before cutting threads. Some die heads have adjustments that allow them to cut threads other than NPT threads. For instance, gas pipe threads have a more pronounced taper built into the entire length of the thread. While this is desirable for a plumber making up a natural gas line, electrical fittings are made for NPT.

Ironworkers and pipefitters may have large bolt dies on a job for threading support rods. These dies make a bolt thread, not an NPT pipe thread. Conduit is always threaded with NPT dies.

**NPT Thread Length.** Couplings and fittings use a straight thread configuration. NPT thread has a taper of ¾″ per foot of thread length so that the conduit and fittings will thread together tightly to form a joint. **See Figure 8-12.** The use of tapered NPT thread on the conduit allows the conduit to engage easily with the coupling or fitting. This allows for easier starts and turns than with other types of threads and also allows for a strong joint to be made once the taper at the shoulder of the thread is engaged in the coupling or fitting.

The length of a standard NPT thread varies depending on the conduit size, from ¾″ for ½″ conduit up to 1¾″ for 6″ conduit. A thread that is cut too short will not fill the coupling or fitting properly and results in the overall conduit length being too long. A thread that is cut too long will bottom out in the coupling or fitting, preventing the tapered shoulder of the conduit thread from engaging the coupling. The coupling is then engaged with the weaker part of the thread, which produces a joint that is not as strong as it should be.

If the thread is too long, the conduit may be turned beyond the halfway point of the coupling. When the next section of conduit is screwed into the coupling, the two sections of conduit will meet near the center. This damages the conduit in the center of the coupling and creates a very sharp edge. Sharp edges can damage conductor insulation and expose the conductors.

It may be necessary to adjust the dies to obtain threads that will properly screw into a fitting. The galvanizing on some fittings may reduce the internal diameter of their threads, causing the electrician to have to cut a deeper thread on the conduit in order to compensate. As a rule, there should be no less than five threads worth of contact between the conduit and the fitting.

**Longer Threads.** There are a few occasions when producing a thread that is longer than standard is desirable. When conduit enters an enclosure that is made of heavy cast material, a longer thread may be necessary to allow for a locknut to be mounted on the outside and a bushing and locknut to be mounted on the inside. Note that the NEC® does not allow the use of longer threads (running thread) with couplings. A standard NPT thread is required with couplings.

> **Tech Tip**
>
> The metal chips, or swarf, produced by threading can be very sharp and should never be handled without gloves. Appropriate PPE should be worn when cutting and threading conduit to prevent metal chips from becoming lodged in the skin or eyes.

**Tech Tip**

A threading machine can be set up with the front end slightly lower than the back to allow the cutting oil to drain out of the conduit, making cleanup easier.

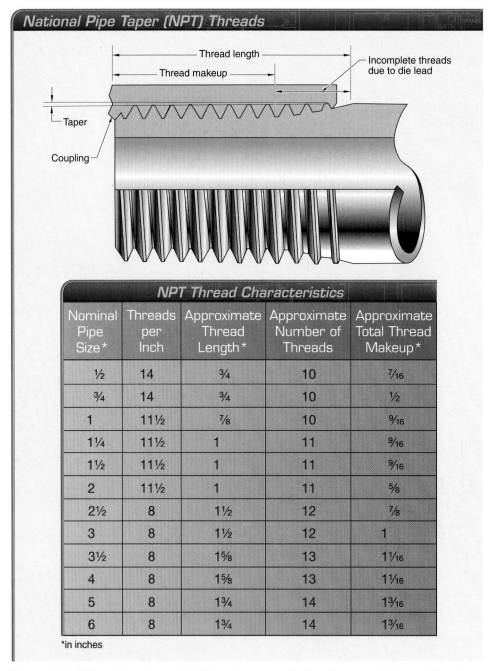

### National Pipe Taper (NPT) Threads

| NPT Thread Characteristics | | | | |
|---|---|---|---|---|
| Nominal Pipe Size* | Threads per Inch | Approximate Thread Length* | Approximate Number of Threads | Approximate Total Thread Makeup* |
| ½ | 14 | ¾ | 10 | ⁷⁄₁₆ |
| ¾ | 14 | ¾ | 10 | ½ |
| 1 | 11½ | ⅞ | 10 | ⁹⁄₁₆ |
| 1¼ | 11½ | 1 | 11 | ⁹⁄₁₆ |
| 1½ | 11½ | 1 | 11 | ⁹⁄₁₆ |
| 2 | 11½ | 1 | 11 | ⅝ |
| 2½ | 8 | 1½ | 12 | ⅞ |
| 3 | 8 | 1½ | 12 | 1 |
| 3½ | 8 | 1⅝ | 13 | 1¹⁄₁₆ |
| 4 | 8 | 1⅝ | 13 | 1¹⁄₁₆ |
| 5 | 8 | 1¾ | 14 | 1³⁄₁₆ |
| 6 | 8 | 1¾ | 14 | 1³⁄₁₆ |

*in inches

**Figure 8-12.** *National Pipe Taper (NPT) threads are standardized to ensure that conduit and couplings engage properly and provide a solid joint.*

## Conduit Threading Procedure

The correct procedures and proper tools must be used to thread conduit and maintain a precision thread. A thread that is improperly cut results in a piece of useless conduit. Conduit is cut and threaded in a threading machine as follows:

1. Secure the conduit in a threading machine by tightening the rear and front chuck. Cut the conduit to the required length and ream the end to remove any burrs or sharp edges. **See Figure 8-13.**

2. Test the threading machine to ensure the conduit is rotating in the correct direction. Oil the conduit and slide the die

head to the end of the conduit. Rotate the lever or mechanism on the die head to engage the dies.

3. Thread the conduit to the correct length while continually lubricating the die. Pressure is not needed on the die head once the threading has started.

4. Open the die head and stop the motor once the proper thread length is reached. Clean the threads with a wire brush or cloth and swab out any excess cutting oil and swarf from the inside of the conduit.

It is important to open the die head before shutting the motor off. This prolongs the life of the dies. A threading machine should not be operated in reverse with the die head engaged. Unlike hand dies, these dies are not made to run in reverse and the cutting edges of the dies will chip or fragment.

When assembling conduit, the coupling should be installed on the end, hand tightened, and also tightened with a wrench. It should take approximately one to two turns with a wrench to completely tighten the assembly.

**Tech Fact**

On 2″ and smaller conduit, a standard thread will be produced when the end of the conduit is flush with the end of the die.

### Threading Conduit

1. Cut conduit to length and ream.

2. Engage dies.

3. Cut threads.

4. Clean threads and inside of conduit.

**Figure 8-13.** *The correct procedures must be followed and the proper tools must be used to properly thread conduit and maintain a precision thread.*

**Threading Large Conduit.** Standard threading machines are typically used to thread conduit up to 2″ trade size, although there are models that can thread conduit up to 4″. Special tools are used to thread larger conduit. Conduit from 2½″ to 6″ can be threaded using a die head that has a chuck and a set bolt so that it can be mounted directly on the conduit. **See Figure 8-14.**

A power pony or a threading machine with a special square attachment or shaft in the drive mechanism can be used to supply power to the equipment. A reduction gear is used to turn the die head on the conduit. **See Figure 8-15.** Many models also feature a clutch that automatically disengages the die head from the power supply if the die head jams.

The die heads for this type of equipment are marked with start and stop marks for a standard thread conduit. The indicator on the die head must be aligned with the start mark before the head is secured to the conduit. The thread is complete and at the proper length when the die head is rotated until it reaches the stop mark. Once the thread is complete, the die head should be opened before reversing the motor.

**Large Conduit Die Heads**

**DIE HEAD**

**MOUNTED ON LARGE CONDUIT**

*Figure 8-14. Larger conduit can be threaded using a die head that has a chuck and a set bolt so that it can be mounted directly on the conduit.*

**POWER PONY**

**THREADING MACHINE**

*Figure 8-15. A power pony or a threading machine with a special square attachment or shaft in the drive mechanism can be used to supply power.*

## SUMMARY

- A die is a cutting tool used to form external screw threads in conduit.

- A die head is the part of a conduit threader that holds the dies securely in position and applies pressure to cut external threads.

- Specifically formulated cutting oils should be liberally applied during the thread cutting process to keep die wear to a minimum.

- A reamer is a tool used to remove sharp edges and burrs from conduit after it has been cut.

- Hand-driven threaders use the arm strength of the electrician to turn the die head on the conduit.

- Power-driven threaders use a reversible electric motor to turn the die head.

- Large power threaders have a fixed die head and rotate the conduit.

- National Pipe Taper (NPT) threads are used on conduit and fittings. An NPT thread has a taper of ¾″ per foot of thread length.

- The length of a standard NPT thread varies depending on the conduit size, from ¾″ for ½″ conduit to 1¾″ for 6″ conduit.

- Special tools are used to thread larger conduit. Conduit from 3″ to 6″ can be threaded using a die head that has a chuck and a set bolt so that it can be mounted directly on the conduit.

# Advanced Bending Techniques

## CONDUIT BENDING and FABRICATION

**9**

SEGMENTED BENDS.................................................................140
CONCENTRIC BENDS.............................................................147
APPLICATION—ARCHED CEILING.......................................152
SUMMARY..................................................................................154

The great majority of conduit bending is done on fixed-radius tools. The bends are made in one shot on a shoe that provides a finished bend with a radius that is defined by the shoe. However, there are certain situations and certain sizes of conduit where one-shot fixed-radius bend techniques are not adequate.

When running conduit on a curved surface, the radius of the surface, not the radius of the shoe, defines the radius of the bend. In addition, there are no readily available benders that will bend 5″ or 6″ rigid conduit in one shot. Segmented bends are used to fabricate both of these types of bends. Segmented bends give complete control over the radius, length, and placement of the bend.

There are four variables used when making segmented bends. These variables are bend radius, completed bend angle, developed length, and number of segments in the bend. A segmented bend can be fabricated to fit around either the outside or inside of a surface. A further development of segmented bends is concentric bends, where several segmented bends are made around a common centerpoint, but each with a different bend radius.

### OBJECTIVES

1. List the variables involved in making segmented bends.
2. Perform the calculations, lay out the conduit, and fabricate segmented bends.
3. Perform the calculations, lay out the conduit, and fabricate concentric bends.

## SEGMENTED BENDS

A *segmented bend* is a bend that consists of a series of small bends made at predetermined locations on a piece of conduit to create one large bend. Each small bend is called a shot. If 10 small bends are used to make one larger bend, it is called a 10-shot bend.

Segmented bends offer complete control over the radius, length, and placement of the bend. Segmented bends require careful calculation and layout for production of acceptable bends. Any error in calculation or placement is repeated over all the bends, resulting in an unacceptable completed bend.

### Segmented Bend Variables

There are four variables used when making segmented bends. These variables are radius, completed bend angle, developed length, and number of segments in the bend. After these variables are determined, the bend can be laid out.

> **Tech Tip**
>
> Many tanks and vessels are insulated. The thickness of the insulation must be included in the radius.

**Bend Radius.** Every bend has a bend radius. For bends made with a one-shot bender, the radius of the shoe determines the bend radius. However, there are many situations where the required bend radius is larger than the radius of the shoe, such as when the radius is fixed by the surface on which the conduit is being mounted. **See Figure 9-1.**

For example, if a conduit run must be bent around a cylindrical storage tank, the required bend radius is determined by the size of the tank. The size of the tank may be shown on mechanical or architectural drawings. If drawings are not available, the radius, diameter, or circumference of the tank needs to be measured.

The simplest method to measure the radius or diameter is to measure across the top of the tank with a tape measure or rule. The diameter is the maximum distance across the top through the center. The radius is exactly half the diameter. If the top is covered or obstructed so the diameter cannot be measured, the next easiest way to determine the radius is to measure the circumference.

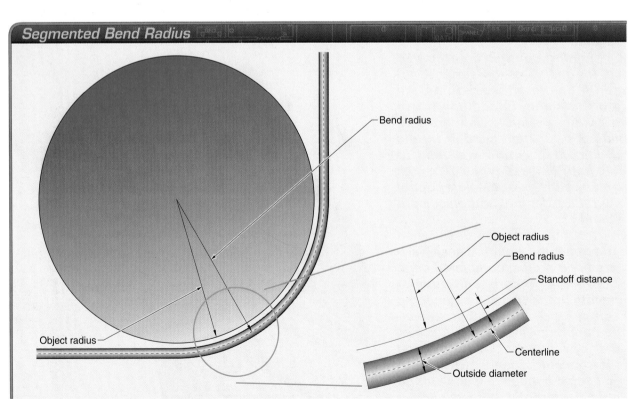

**Figure 9-1.** *Every bend has a bend radius that is slightly larger than the radius of the object around which it is bent.*

The circumference is equal to the length of a string or conductor that wraps around the tank. The radius is the circumference divided by 6.28 ($R = C \div 2\pi$). **See Appendix.**

The bend radius is slightly larger than the radius of the object. Once the radius of the circular object is determined, allowances must be made for spacing between the conduit and the surface, and for size, or the outside diameter (OD), of the conduit. The bend radius is the distance from the center of the circular object to the centerline of the conduit as it bends around the object.

For example, 1″ rigid conduit is being bent around an object that has a 36″ radius. The OD of this conduit is 1.315″. Half the OD must be added to the radius of the object to determine the centerline of the conduit bend. Half this OD is about 0.66″.

A distance must be chosen for the spacing between the conduit and the surface. This distance depends on the mounting method. If the conduit run is fairly short and no support is needed, or if the conduit can be mounted flat against the surface, the conduit can be run tightly against the surface and the total bend radius will be 36.66″ (36 + 0.66 = 36.66). If a standoff holds the conduit 1″ away from the surface, that 1″ must be added to the radius. The total bend radius is then 37.66″ (36 + 1 + 0.66 = 37.66).

**Completed Bend Angle.** The completed bend angle depends on the distance that the conduit run must travel around the circular object. This distance will be a fraction of the circumference of the object. **See Figure 9-2.** For example, a circle has 360°. If a conduit run must travel all the way around the object, the bend angle is 360°. If a conduit run must travel one-fourth of the way around the object, the bend angle is 90° (360° ÷ ¼ = 90°).

**Developed Length.** The developed length is the amount of conduit used in a bend. **See Figure 9-3.** The developed length is determined by the field conditions and can vary from a small section of conduit to an entire 10′ length or more.

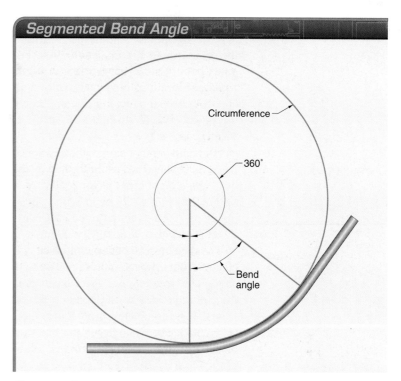

**Figure 9-2.** *The completed bend angle depends on the fraction of the distance that the conduit run must travel around the circular object.*

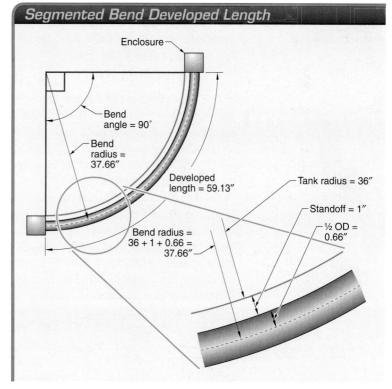

**Figure 9-3.** *The developed length is the amount of conduit used in a bend and can vary from a small section of conduit to an entire 10′ length.*

A common application for segmented bends is to run the conduit one-fourth of the way around the object, or 90°. For a 90° bend, the developed length is the bend radius multiplied by the factor 1.57 ($L = 1.57R$). For a bend radius of 37.66″, the developed length is 59.13″ ($1.57 \times 37.66 = 59.13$).

For bends at other angles, the developed length is the bend radius multiplied by the bend angle and by the factor 0.0175 ($L = 0.0175R\theta$). For a 45° bend with a bend radius of 37.66″, the developed length is 29.66″ ($0.0175 \times 37.66 \times 45 = 29.66$).

The developed length can also be calculated from the bend radius and the bend angle. The bend radius is used to calculate the circumference of the conduit as if it were to bend completely around the object ($C = 2\pi R$). For example, if the bend radius is 37.66″, the circumference is 236.6″ ($2\pi \times 37.66 = 236.6$).

The bend angle is used to calculate the fraction of the distance that the conduit travels around the object. For example, if the bend angle is 60°, the conduit travels one-sixth of the way around the object ($60 \div 360 = \frac{1}{6}$). The developed length is the product of the circumference and the fraction of the distance. In this example, the developed length is 39.43″ ($\frac{1}{6} \times 236.6 = 39.43$).

**Number of Segmented Bends.** Once the completed bend angle and the developed length are known, the number and precise angle of the segmented bends can be chosen. There is no exact method for choosing the number of segments.

If many shots are used, each angle will be small and the individual bends will be less noticeable. This allows more opportunities to correct small errors in earlier bends so that the final bend looks correct. However, more bends also means more fabrication time and more opportunity for error.

If fewer shots are used, each angle will be larger and more noticeable. The segmented bend will look choppy, but fewer bends means less fabrication time. A common rule of thumb for best appearance states that the individual bends should be no more than 6° and no more than 4″ apart.

When choosing the number of shots to be used when fabricating 90° bends, there are several choices that produce round degree values. This can make it easier to bend accurate angles. For example, a 90° bend made with 18 shots requires 5° bends ($90 \div 18 = 5$). The same 90° bend made with 9 shots requires 10° bends. However, using only 9 shots will result in a choppy looking bend. **See Figure 9-4.**

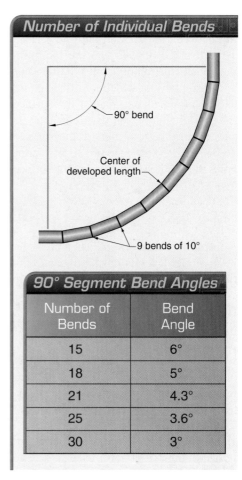

*Number of Individual Bends*

90° bend

Center of developed length

9 bends of 10°

| *90° Segment Bend Angles* | |
|---|---|
| Number of Bends | Bend Angle |
| 15 | 6° |
| 18 | 5° |
| 21 | 4.3° |
| 25 | 3.6° |
| 30 | 3° |

**Figure 9-4.** *Once the completed bend angle and the developed length are known, the number and precise angle of the individual bends can be chosen.*

## Segmented Bend Layout

Once the developed length and the number of segmented bends are known, the conduit can be marked for the bends. Depending on the field conditions, the conduit may be laid out so the segmented bend begins some distance from the end to allow a straight section to reach a desired point.

The distance between shots is the developed length divided by the number of shots. The distance between shots is calculated as follows:

$$D = \frac{L}{N}$$

where

$D$ = distance between shots
$L$ = developed length
$N$ = number of shots

For example, a 90° bend is to be made with 21 shots of 4.3° each. **See Figure 9-5.** The developed length is 72″. The distance between shots is calculated as follows:

$$D = \frac{L}{N}$$

$$D = \frac{72}{21}$$

$$D = \mathbf{3.43''}, \text{ or } \mathbf{3\frac{7}{16}''}$$

**Odd Number of Shots.** The easiest method of laying out segmented bends is to use an odd number of shots. To use this method, the ends of the developed length must be laid out on the conduit with temporary pencil marks. The center of the developed length is marked midway between the ends. **See Figure 9-6.** All bend marks are then laid out on either side of the center pencil mark. Commonly, the developed length will be laid out near the center of the piece of conduit. This leaves a length on each end to be able to start the conduit in the bender. This length depends on the bender used.

For example, for a 90° bend with a developed length of 72″ and 21 shots of 4.3°, the distance between shots is 3⁷⁄₁₆″ (72 ÷ 21 = 3.43). Two temporary marks are placed 72″ apart on the conduit, leaving enough room on each end for handling the conduit. A pencil mark is placed exactly halfway between the two temporary marks, leaving 20 more pencil marks to place.

Ten additional pencil marks are placed on either side of the center pencil mark, spaced 3⁷⁄₁₆″ apart. The conduit ends up with one pencil mark in the exact center of the developed length and 10 pencil marks on each side of the center. The two outside bend marks will end up just inside the temporary marks.

The 4.3° bends are centered on the pencil marks. For one-shot benders, the center of a 4.3° bend should be found and placed on the bender shoe. For a multiple-shot hydraulic bender, the bends should be made on the center of the shoe. A no-dog device should be used to keep all the individual bends in the same plane, as segmented bends are particularly subject to doglegs.

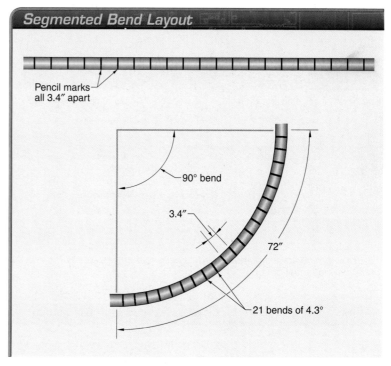

**Figure 9-5.** *Once the developed length and the number of segmented bends are known, the conduit can be marked for the bends.*

**Even Number of Shots.** An even number of shots may be chosen in order to end up with individual bends at a round bend angle. For example, for a 90° bend with 18 shots, the individual bends are 5° (90 ÷ 18 = 5). With a developed length of 72″, the distance between shots is 4″ (72 ÷ 18 = 4). As with an odd number of shots, two temporary marks are placed 72″ apart on the conduit, leaving enough room on each end for handling the conduit. **See Figure 9-7.**

When an even number of shots is used, the bend layout proceeds in a similar manner as with an odd number of shots. Since there is an even number of shots, there is no bend at the exact center of the developed length. The two center bend pencil marks are placed equidistant from the center of the developed length, spaced 4″ apart, leaving 16 more pencil marks to place. The remaining pencil marks are placed 4″ apart, with eight on each side of the center.

The layout ends up with nine pencil marks on each side of the center. Again, the bends are centered on the pencil marks. A bend is not made at the exact center, since the bend marks were placed on both sides of the center.

## Application—Enclosures Mounted on Pressure Vessel

Two instrumentation enclosures are mounted exactly 90° apart on a pressure vessel. A run of ¾″ rigid conduit connecting the two enclosures is being placed around the vessel. The vessel measures 10′-1¼″ (121¼″) in diameter. The conduit is to be mounted tight to the vessel surface. **See Figure 9-8.**

The OD of ¾″ rigid is 1.050″. Half of this is 0.525″. Since the pressure vessel diameter is 121¼″, the vessel radius is 60⅝″ (121¼ ÷ 2 = 60⅝). The bend radius is 61⅛″ (60⅝ + 0.525 = 61.15). Since the bend is 90°, the developed length is 1.57 times the bend radius. In this case, the developed length is 96″ (1.57 × 61.15 = 96).

Because the work is exposed, the developed length is relatively long, and the conduit must fit tight to the vessel surface, a 90° bend with 25 shots of 3.6° is chosen. The developed length is 96″ and the distance between shots is 3¹³⁄₁₆″ (96 ÷ 25 = 3.84). A piece of conduit can be laid out with one mark at the center of the developed length and 12 marks on each side of the center, 3¹³⁄₁₆″ apart. These marks serve as the center of the bends for the 25 individual bends.

A good bender for this type job is an electric bender. Electric benders can be set up to fabricate accurate repeatable bends. The only preparation will be fabricating a 3.6° bend on a scrap piece of conduit, finding the center of the bend, and transferring that mark to the bender shoe. This mark is the benchmark for all 25 individual 3.6° bends.

Segmented bends can also be fabricated on other types of benders. When using these benders, a high-quality bending protractor should be used to ensure that the segmented bends are all the same. As with the other benders, the center of the bend should be found, transferred to the bending shoe, and used as a benchmark. This will make certain the segmented bend begins and ends at the correct location.

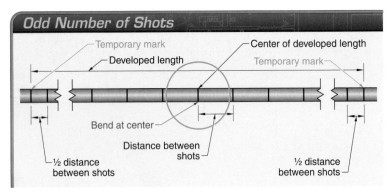

**Figure 9-6.** *With an odd number of shots, a bend mark is placed at the center of the developed length.*

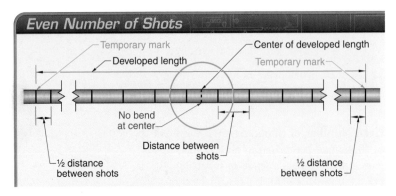

**Figure 9-7.** *With an even number of shots, two bend marks are placed an equal distance on either side of the center of the developed length.*

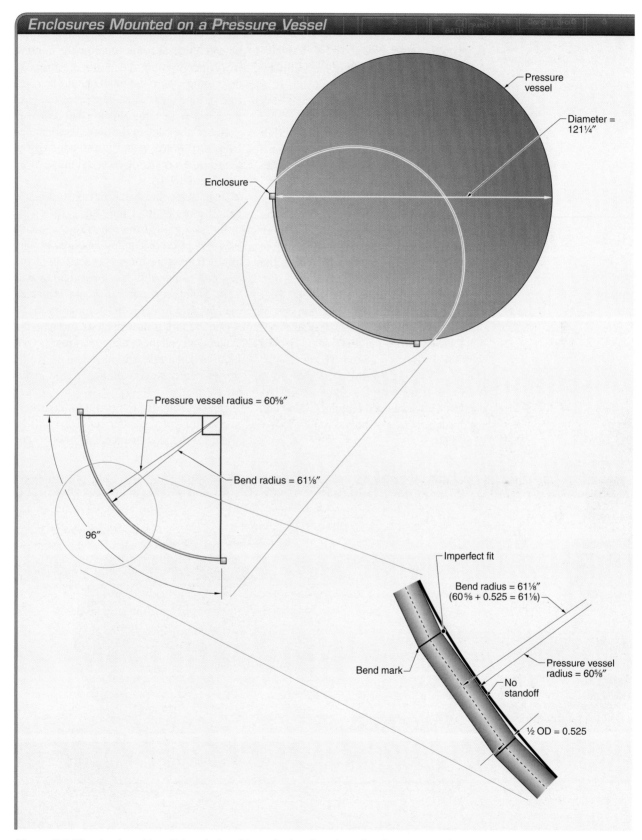

**Enclosures Mounted on a Pressure Vessel**

Pressure vessel

Diameter = 121¼″

Enclosure

Pressure vessel radius = 60⅝″

Bend radius = 61⅛″

96″

Imperfect fit

Bend radius = 61⅛″
(60⅝ + 0.525 = 61⅛)

Pressure vessel
radius = 60⅝″

Bend mark

No
standoff

½ OD = 0.525

*Figure 9-8. The developed length is calculated from the bend angle and radius.*

### Application—Control Boxes Mounted near Cylindrical Heating Tank

A large manufacturing plant has cylindrical tanks that are used to mix and heat materials. The control box for each tank is bracket-mounted near each unit and fed with a 3″ rigid conduit. The feeder conduit follows the contour of the tank for 90° on a rack on its way to the switchgear room. The diameter of the tank is measured at 7′-2⅜″ (86⅜″). **See Figure 9-9.**

Since the tank diameter is 86⅜″, the tank radius is 43³⁄₁₆″ (86⅜ ÷ 2 = 43³⁄₁₆). Because the tank is heated, a 3¼″ double unistrut® is used as a standoff to keep the conduit away from the hot surface.

The OD of 3″ rigid conduit is 3.50″. Half of this is 1¾″. The bend radius is the sum of the tank radius, the standoff, and half the conduit OD. Therefore, the bend radius is 48³⁄₁₆ (43³⁄₁₆ + 3¼ + 1¾ = 48³⁄₁₆). The developed length is 75⅝″ (1.57 × 48³⁄₁₆ = 75.65).

A 90° bend with 21 shots of 4.3° is chosen. The developed length is 75⅝″ and the distance between shots is 3.60″ (75⅝ ÷ 21 = 3.60).

The control panel is not mounted directly on the tank. A straight distance must be taken into account to allow the conduit to properly enter the control panel. To determine where the bend should begin, a rule or tape measure can be held square to the edge of the control panel and touching (tangent) the edge of the tank. The distance from the control panel to the edge of the tank is the length of the straight section. **See Figure 9-10.**

In this case, the distance from the control panel to the edge of the tank is 36″. This is the straight distance before the bends begin. A piece of conduit can now be laid out. A temporary mark is placed 36″ from one end. Since the developed length was 75⅝″, another temporary mark is placed that distance from the first temporary mark. A pencil mark is now placed exactly in the center between these temporary marks. Ten more pencil marks are placed on each side of the center pencil mark, spaced 3.60″ (3⅝″) apart.

The spacing from the ram to the rollers in the front of a bender typically will be about 18″, depending on bender model. The

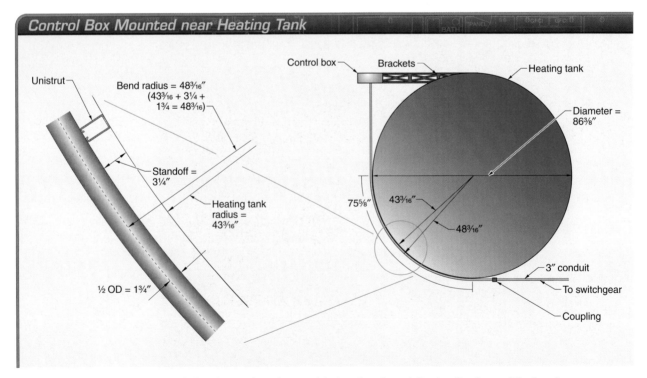

**Figure 9-9.** *A layout for a concentric bend sometimes has straight lengths of conduit extending beyond the bend.*

minimum spacing from the ram to the vise in the back is about the same distance or a little more. There is a 36″ straight section at the beginning of the conduit, and the developed length is 75⅝″. The actual length of the threaded rigid conduit is 118¾″. This leaves 7¼″ of straight conduit after the bends.

This layout will not fit in the bender because of the short distance from the end to the first bend. A nipple and coupling can be added to that end to provide enough length to properly fit in the bender. However, the shoe should never actually bend directly on the coupling because it will likely damage the threads and possibly the bender, too.

Because the conduit for this segmented bend is fairly large, it will need to be bent with a hydraulic bender. There are multishot hydraulic benders available that use short shoes to efficiently bend small angles, and accurate angles can be fabricated with the use of a high-quality bending protractor.

One-shot benders can also be used for segmented bends. For best results, the center of the bend should be found on a scrap conduit and transferred to the bender shoe so that the bend is properly located.

## CONCENTRIC BENDS

A *concentric bend* consists of two or more segmented bends nested together. Concentric bends all have the same centerpoint, but each has a different radius. Each successive piece of conduit has a different radius and developed length than the original conduit. **See Figure 9-11.**

The first bend is made by the same method as any standard segmented bend. Other lengths of conduit are bent with a larger or smaller radius. The change in radius is equal to the size of the conduits plus the spacing between the conduits. Once the new radius is calculated, the developed length of the next conduit is calculated using the new radius. Once these calculations are made, each conduit can be laid out and bent in segments.

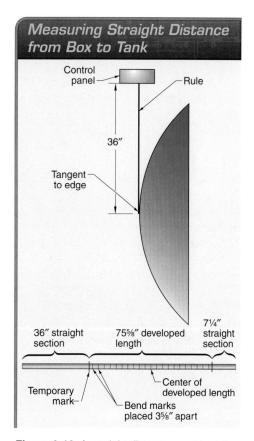

**Figure 9-10.** *A straight distance must be taken into account to allow the conduit to properly enter the control panel.*

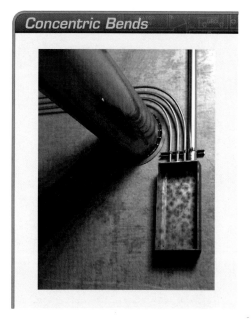

**Figure 9-11.** *Concentric bends all have the same center point, but have different radii. Each successive section of conduit has a different radius and developed length than the original bend.*

## Application—Conduit Rack with Concentric Bends

Concentric bends can be challenging to lay out and fabricate, and it is helpful to sketch out the installation and determine all of the measurements and calculations before attempting to bend the conduits. Before the various calculations are made, the outer diameters of the various sizes of conduits must be known and the center-line radius of the reference conduit must be determined.

For example, a rack of rigid conduit with concentric 90° bends is to be run out of a junction box around a circular air duct. The inside conduit is 2″ (conduit A), the middle conduit is 2½″ (conduit B), and the outside conduit is 3″ (conduit C). The conduits are to be placed with equal spacing of 3½″ between them. **See Figure 9-12.**

The duct has a radius of 28¹³⁄₁₆″ and the inside conduit is to be run tight against the duct. The OD is 2.375″ for conduit A, and half the OD is 1³⁄₁₆″ (2.375 ÷ 2 = 1.1875). Therefore, the bend radius of conduit A is 30″ (28¹³⁄₁₆ + 1³⁄₁₆ = 30). **See Figure 9-13.**

For the remaining conduits, the OD is 2.875″ for conduit B, and 3.500″ for conduit C. Half the OD of conduit B is 1⁷⁄₁₆″ (2.875 ÷ 2 = 1.4375) and half the OD of conduit C is 1¾″ (3.500 ÷ 2 = 1.750).

The bend radius of conduit B is the sum of the bend radius of conduit A, half the OD of conduit A, the spacing between them, and half the OD of conduit B. Each radius is calculated relative to another, in sequence, as follows:

$$R_B = R_A + \frac{1}{2}OD_A + S + \frac{1}{2}OD_B$$

where

$R_B$ = bend radius of conduit B, in inches

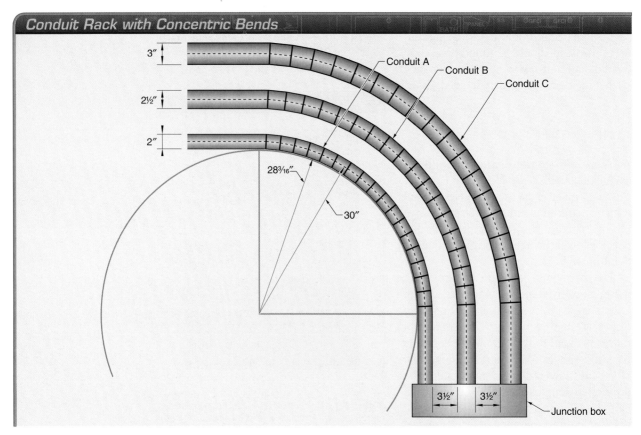

**Conduit Rack with Concentric Bends**

3″  
2½″  
2″  

28¹³⁄₁₆″  
30″  

Conduit A  
Conduit B  
Conduit C  

3½″  3½″  
Junction box

*Figure 9-12. Parallel conduit runs can be placed with even spacing between the conduits.*

$R_A$ = bend radius of conduit A, in inches

$OD_A$ = outside diameter of conduit A, in inches

$S$ = spacing between conduits, in inches

$OD_B$ = outside diameter of conduit B, in inches

Therefore, the bend radii of conduit B and conduit C are calculated as follows:

$$R_B = R_A + \frac{1}{2}OD_A + S + \frac{1}{2}OD_B$$

$$R_B = 30 + 1.1875 + 3.5 + 1.4375$$

$$R_B = \textbf{36.125''}, \text{ or } \textbf{36⅛''}$$

$$R_C = R_B + \frac{1}{2}OD_B + S + \frac{1}{2}OD_C$$

$$R_C = 36.125 + 1.4375 + 3.5 + 1.75$$

$$R_C = \textbf{42.8125''}, \text{ or } \textbf{42¹³⁄₁₆''}$$

The next step is to calculate the developed lengths of the three conduits and determine the number of shots that should be used to make the three bends. Since the conduits are all bent to 90°, the developed length is the product of the constant 1.57 and the bend radius. The developed lengths are calculated as follows:

$$L = 1.57 \times R$$

where

$L$ = developed length of the conduit, in inches

$R$ = bend radius of the conduit, in inches

Subscripts can be used to identify the conduit being used in the calculations. The three developed lengths are calculated as follows:

$$L_A = 1.57 \times R_A$$
$$L_A = 1.57 \times 30$$
$$L_A = \textbf{47.1''}, \text{ or } \textbf{47⅛''}$$

$$L_B = 1.57 \times R_B$$
$$L_B = 1.57 \times 36.125$$
$$L_B = \textbf{56.72''}, \text{ or } \textbf{56¾''}$$

$$L_C = 1.57 \times R_C$$
$$L_C = 1.57 \times 42.8125$$
$$L_C = \textbf{67.22''}, \text{ or } \textbf{67¼''}$$

| Concentric Bend Dimensions | | | | | |
|---|---|---|---|---|---|
| Trade Size* | OD* | Half OD* | Bend Radius* | Distance between Shots* | Developed Length* |
| 2 | 2.375 | 1³⁄₁₆ | 30 | 2⅝ | 47⅛ |
| 2½ | 2.875 | 1⁷⁄₁₆ | 36⅛ | 3⅛ | 56¾ |
| 3 | 3.500 | 1¾ | 42¹³⁄₁₆ | 3¾ | 67¼ |

*in inches

**Figure 9-13.** *A table of dimensions can be used to simplify layout calculations for concentric segmented bends.*

Since conduit C has the longest developed length, it should be used to determine the number of shots that will be used for all three conduits. This is because it is also necessary to evaluate the distance between bends as part of the layout procedure. For example, if this bend is to be made with 15 shots of 6° each, the distance between bends for conduit C will be 4.48" (67.22 ÷ 15 = 4.48), slightly over the recommended maximum of 4".

If the same bend is made with 18 shots of 5° degrees, the distance between bends will be 3.73" (67.22 ÷ 18 = 3.73), which will comply with the rule of thumb of not more than 6° or 4". Therefore, all three conduits should be fabricated using 18 shots of 5°.

The distance between shots on conduit A is 2.61" (47.1 ÷ 18 = 2.62). The distance between shots for conduit B is 3.15" (56.72 ÷ 18 = 3.15). While conduits A and B could be made with fewer shots, the end result will look better if the same number of shots is used for all of the conduits.

It should be kept in mind that when making segmented bends the radius is measured to the center of the conduit. This can create confusion because measurements for stubs are generally made to the back, not the center of the bend. The overall length of the stub or leg of the conduit for a segmented bend is the length of the straight conduit, plus the radius of the bend, plus ½ of the OD of the conduit.

The distance from the junction box to the beginning of the bend is measured to be 17⁷⁄₁₆″. **See Figure 9-14.** It is necessary to add about 1″ to the overall length of the stub in order to have enough thread inside the box to accommodate the locknut and bushing. Therefore, the straight length measured from the end of the threads inside the junction box to where the developed length begins is 18⁷⁄₁₆″. This distance is the same for all three conduits.

To determine how much straight conduit there will be on the horizontal leg of conduit C, the developed length plus the length of the straight conduit before the bend is subtracted from the overall length of the conduit. The length of the 3″ conduit, without the coupling, is 118¾″. The length of the conduit which forms the horizontal leg of the bend is therefore 33⅛″ (118¾ – 67.22 – 18⁷⁄₁₆ = 33.1).

It must be decided during the layout process if the other two conduits will be fabricated so that the ends line up with the end of conduit C. It may not always be necessary or desirable to make the end of the conduits line up with each other. However, for exposed work the final product will look more professional if the ends are aligned.

Lining up the ends may also provide the advantage of simplifying the installation of the remaining conduit run. By having the couplings in the same vicinity for all of the conduits, fewer moves will be needed during the installation. This can be an important factor at times when ladders are used to get the worker up to the conduit run.

In order to line up the ends, conduit A and conduit B will have to be cut and threaded before bending begins. Since the amount of straight pipe to the left of the

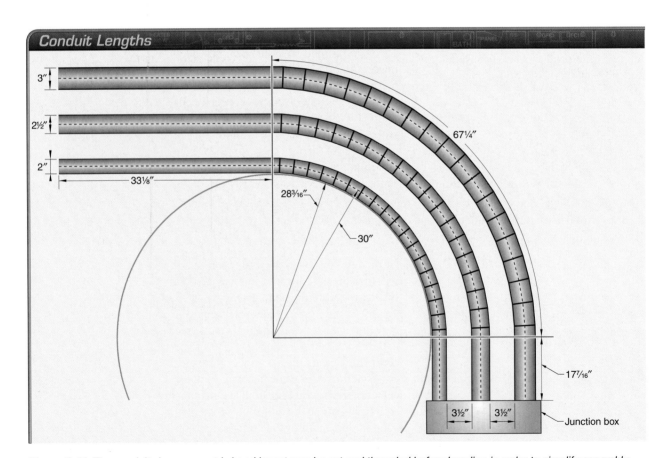

**Figure 9-14.** *The conduits in a concentric bend layout may be cut and threaded before bending in order to simplify assembly.*

developed length will be same for all three conduits, the lengths of conduit A and conduit B can be calculated by simply adding the amount of straight conduit required at both ends to the developed length of each conduit. **See Figure 9-15.**

Therefore, conduit B will be cut and threaded to a length of 108¼″ (33.1 + 56.72 + 18.4375 = 108.26). Conduit A will be cut and threaded to a length of 98⅝″ (33.1 + 47.1 + 18.4375 = 98.64). Both conduit A and conduit B need to be threaded after cutting. It may be necessary to run a slightly longer thread on the junction box end of conduit A to ensure there is ample thread for locknuts and a bushing. This is because the threads are shorter on the 2″ conduit than on the others. It is also a good idea to check the fit of the locknuts on the factory-cut threads. The galvanized coating sometimes is thicker than expected. If the locknuts do not spin on fully, the threads may have to be chased by running a threading die over them.

Once the conduits have been cut, reamed, and threaded they can be marked for bending. Since all three conduits will be bent with the same amount of straight conduit remaining on either end, they can be laid out from either direction. In this case, the marking will start at the end that will later go into the junction box. Measuring back from the end of each conduit by 18⁷⁄₁₆″, a temporary mark is made to indicate the beginning point of the developed length. **See Figure 9-16.**

Conduit A is then marked with another temporary mark at 47⅛″ from the first mark. This marks the end of the developed length. The center of the developed length is found, and nine marks are made on either side of the center mark, spaced 2⅝″ apart from each other. The spacing from the center mark to the bending marks on either side of it will be half of the distance of a shot, or about 1⁵⁄₁₆″.

Conduit B is then marked with another temporary mark at 56¾″ from the first mark. This marks the end of the developed length. The center of the developed length is found, and nine marks are made on either side of the center mark, spaced 3³⁄₃₂″ apart

from each other. The spacing from the center mark to the bending marks on either side of it will be half of the distance of a shot, or about 1⁹⁄₁₆″.

Conduit C is then marked with another temporary mark at 67³⁄₁₆″ from the first mark. This marks the end of the developed length. The center of the developed length is found, and nine marks are made on either side of the center mark, spaced 3¾″ apart from each other. The spacing from the center mark to the bending marks on either side of it will be half of the distance of a shot, or about 1⅞″.

| Conduit Layout Lengths | | | | | |
|---|---|---|---|---|---|
| Conduit | Trade Size* | Straight Length at Box* | Developed Length* | Straight Length at End* | Total Length* |
| A | 2 | 18⁷⁄₁₆ | 47.10 | 33.10 | 98⅝ |
| B | 2½ | 18⁷⁄₁₆ | 56.72 | 33.10 | 108¼ |
| C | 3 | 18⁷⁄₁₆ | 67.22 | 33.10 | 118¾ |

*in inches

**Figure 9-15.** *The conduit lengths can be placed in a table to help organize the cutting and threading operation.*

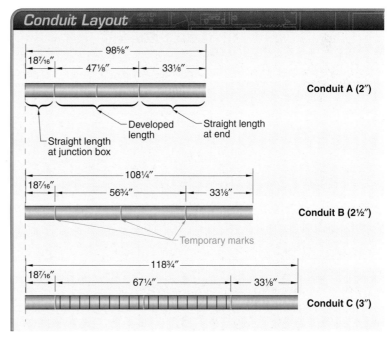

**Figure 9-16.** *When laying out the conduit for cutting and threading, the straight lengths are the same at each end, while the developed length varies from one conduit to another.*

At this point it is a good practice to erase the beginning, end, and center marks for the developed length to avoid possible confusion while bending the conduit. It may also be helpful to number the individual shots to help to keep track of the progress of the bend angle.

## APPLICATION— ARCHED CEILING

A gallery has a circular arched ceiling with arched structural framing members. A run of 2″ EMT must be run across the structure following the contour of the roof members. The EMT will be run tight against the surface. The plans show the ceiling has a 65′ radius. **See Figure 9-17.**

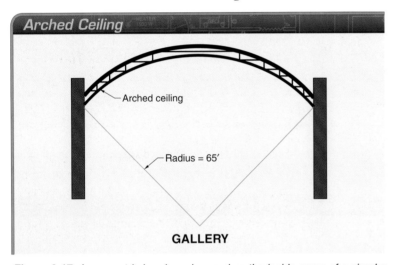

**Arched Ceiling**

Arched ceiling

Radius = 65′

**GALLERY**

**Figure 9-17.** *A concentric bend can be used on the inside curve of a circular surface.*

The OD of 2″ EMT is 2.197″. Half of this is 1.10″ (0.092′). Since the arch radius is 65′ and the conduit is very small relative to the radius, the small amount contributed by the OD of the EMT can be ignored. The circumference where the conduit runs is 408′ ($2 \times \pi \times 65 = 408$).

The bend angle and developed length cannot readily be determined. Another approach is to calculate the degrees of bend per foot of conduit. Since there are 360° in a circle and the circumference is 408′, the constant is 0.882° per foot (360 ÷ 408 = 0.882). In other words, every foot of conduit must bend 0.882°. Therefore, a 10′ length of conduit must bend 8.82°.

It would be very difficult to make 10 bends of 0.882° in a 10′ length of conduit. One bend of 8.82° in the center would look out of place and the conduit would not fit tight against the arch. It is relatively easy to make two 4.5° bends or three 3° bends in a 10′ length of conduit. Either of these types of segmented bends will fit fairly close to the arch. More bends will result in a better fit, but will take longer to fabricate.

---

*T*ech Fact

The NEC® lists the required bend radius for field bends. The radius of the bend is measured to the centerline of the conduit or tubing.

## Tips for Making Segmented Bends . . .

Segmented bends can be fairly complex and difficult to fabricate. There are a few important tips that can make it easier to fabricate accurate segmented bends. These tips are as follows:

- The conduit must not be allowed to rotate during the bending process since segmented bends are particularly susceptible to doglegs. The use of a no-dog level is essential to successful fabrication of a segmented bend. There are many bends that must be kept in the same plane and this can only be done with a no-dog level.

- The progress of the overall bend must be checked with a protractor level or an angle gauge after each shot is made. If the amount of bend produced for a given shot is too large or too small, gradual corrections can be made in subsequent shots to fix the problem.

- The overall bend must be checked for overall angle and for stub length when most of the individual bends have been made and only a few shots remain. It is entirely possible to form a bend that has the proper angle but the wrong length for the stub and the leg.

When the bend has progressed so that there are only three to five shots left to be made, the conduit should be removed from the bender and two measurements taken. First, the overall angle of the bend must be checked. If the individual bends have been properly fabricated the angle should be very close to the desired value for the number of shots completed. If there is too much or not enough bend, it is possible to compensate by slightly over-bending or under-bending the remaining shots.

The second thing that must be checked is the height of the stub. As a rule, if the bend has been properly made, the length of the remaining rise needed will be slightly less than the remaining distance between the remaining shots.

For example, a 90° bend is being fabricated around a storage tank with a 48″ bend radius and a desired stub length of 62″. The developed length is 75.36″ (1.57 × 48 = 75.36). A segmented bend with 21 shots of 4.3° each is chosen. The distance between shots is 3.59″ (75.36 ÷ 21 = 3.59).

After shot 18, the conduit is removed from the bender and the angle is measured at 76.5°, just slightly less than the desired angle of 77.4° (18 × 4.3 = 77.4). The last 3 shots need to bend the conduit by 13.5° (90 − 76.5 = 13.5), or about 4.5° per shot.

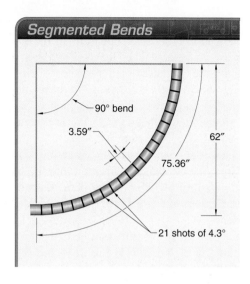

**Segmented Bends**

90° bend

3.59″

75.36″

62″

21 shots of 4.3°

Measuring from the back side of the leg to the end of the stub at its center, it is found that the present height of the stub is 51½″. The remaining amount of rise is therefore 10½″ (62 − 51½ = 10½). The remaining distance between shot 18 and shot 21 is 10 ¾″ (3.59 × 3 = 10.77).

Minor corrections will be needed for the remaining shots since the values for the remaining amount of angle and the rise are close to, but not exactly the same as, the projected values. The conduit is reinserted into the bender and the bender is advanced until the conduit is held snug and aligned with the bending mark for shot 19.

The stub is checked side-to-side to make sure that it is plumb, or level, with the bender. The protractor level is re-zeroed or checked on the leg of the bend. Shot 19 is made at an angle slightly larger than the expected 4.3°, attempting to correct for the slight lack of overall bend. The same procedure is followed for shot 20. The conduit is again removed from the bender and the angle and stub length are measured.

For this example, it is found that the overall angle of bend is 86° and the stub is 59″ long. This means that the amount of angle needed to complete the last bend is approximately the anticipated amount of 4.3°. However, if the last bend is made using the mark for shot 21, without correction, the stub will likely turn out to be about 62 ⅝″ (59 + 3.59 = 62.59), about ⅝″ too long. This happens because the last shot moves the end of the conduit almost vertically, causing the end of the stub to rise by an amount equal to the distance between shots.

### . . . Tips for Making Segmented Bends

In many cases it will not be necessary to adjust the last bend in order to compensate for the discrepancy between the actual and the desired height of the stub, since the conduit will not be exposed. However for some jobs it will be necessary to make sure that the length of the stub is exactly as planned. In these cases, the bending procedure for the last bend must be modified.

To make the bend so that the length of the stub is as desired, the conduit is reinserted into the bender and prepared for bending. It will be necessary to be able to measure the rise in the stub as the last shot is made.

Typically a tape measure can be extended down from the end of the stub to the floor and the stub measurement noted. If the bender is in a horizontal position, a straightedge can be clamped to the leg of the conduit for this measurement. The remaining amount of rise needed is added to the stub measurement to determine when to stop bending.

For this example, the end of the stub is 88" above the floor and needs to rise by 3″ to achieve the 62″ stub. The conduit will be bent until the reading on the tape measure is 91″ (88 + 3 = 91), plus a small amount for springback. At this point the bend will be stopped, the conduit removed from the bender, and the measurements rechecked.

The stub is now the correct length, but the bend is slightly open, by about 1°. To adjust the angle without increasing the length of the stub, the conduit is put back into the bender and 1° of additional bend is added to shot 1 at the opposite end of the conduit. This closes the bend so that it is a true 90° sweep, and does not add additional height to the stub.

This procedure, though time-consuming, is necessary when the utmost in accuracy is needed for a segmented 90° bend. The extra time spent measuring and correcting the nearly complete bend is often the difference between a bend that fits and one that must be sent to the scrap pile.

## SUMMARY

- A segmented bend is a bend that consists of a series of small bends made at predetermined locations on a piece of conduit to create one large bend.

- The four variables used to make segmented bends are bend radius, completed bend angle, developed length, and number of segments in the bend.

- For segmented bends, the required bend radius is larger than the radius of the shoe, such as when the radius is fixed by the surface on which the conduit is being mounted.

- The completed bend angle depends on the distance that the conduit run must travel around the circular object. This distance will be a fraction of the circumference of the object.

- The developed length is the amount of conduit used in a bend and can vary from a small section of conduit to an entire 10′ length.

- There are several choices of bend angle that produce even degree values that make layout and bending fairly easy.

- Concentric bends consist of two or more segmented bends nested together where all have the same center point, but each has a different radius.

# Underground Conduit Installation Procedures

## CONDUIT BENDING and FABRICATION

**10**

JOB PLANNING .................................................................. 156
SURVEYING INSTRUMENTS................................................ 158
EXCAVATION..................................................................... 164
UNDERGROUND INSTALLATION PROCEDURES.................... 169
BACKFILLING ................................................................... 171
SUMMARY ........................................................................ 174

*A job specification typically contains a site plan, mechanical drawings, and electrical drawings that identify the requirements of the job. Any variance from the specifications must be approved before beginning work on that section.*

*A benchmark is a monument or mark placed by a surveyor that shows precise location and elevation. Surveyors may place corner stakes, known as hubs, at each corner of a property. Hubs form control lines that will be used to lay out light pole bases, utility holes, conduit runs, and any other electrical work in that area.*

*A builder's level is an instrument consisting of a telescope with crosshairs and a spirit level. A builder's level has one plane of motion. A transit has a telescope that rotates vertically and horizontally.*

*All employees entering a confined space must be instructed as to the nature of the hazards, necessary precautions, and protective and emergency equipment required. Sloping is the process of cutting back trench walls to an angle that eliminates the chance of collapse. Shielding is the use of a portable protective device capable of withstanding forces from a cave-in.*

## OBJECTIVES

1. Explain how to use benchmarks and hubs.
2. Describe the difference between a builder's level and a transit.
3. List six safety rules to follow to prevent injury from cave-ins.
4. Describe how sloping, shoring, and shielding are used to protect workers in trenches.
5. Explain why different types of concrete are used.
6. Explain why compaction is an important part of backfilling an excavation.

## JOB PLANNING

In order to complete a job efficiently, the job must be planned thoroughly before beginning work. Planning ensures that the labor, tools, and materials are in place and that any information needed to do the job is available. The job specifications cover the scope of work and allow for a planned strategy.

### Job Specifications and Site Plans

The amount of information in a job specification and site plan depends on the size and complexity of the job. Drawings are provided separately. **See Figure 10-1.** The job specifications should be reviewed because the specifications are the requirements of the job. Any variance from the specifications must be approved before beginning work on that section.

The type of utility holes, type and schedule of conduit, and any spacing or depth requirements greater than required by the NEC® should be found in the electrical section of the specifications. The section may also include a materials list.

The electrical section may contain concrete encasement provisions. The general contract and excavation sections of the specifications must also be consulted. The backfill material and compaction requirements for all excavations are defined in these sections.

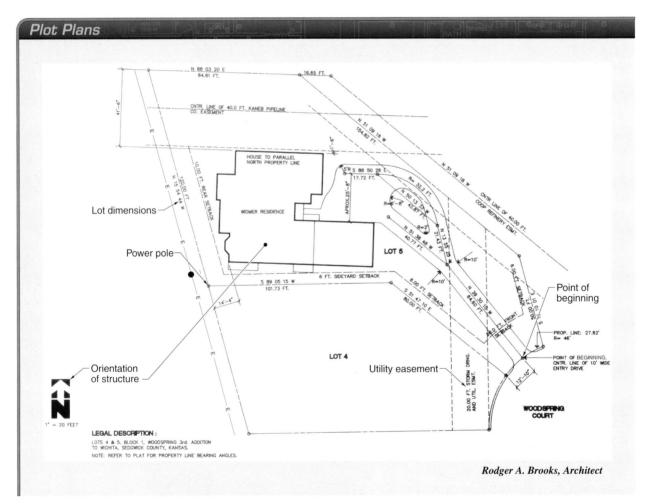

*Rodger A. Brooks, Architect*

**Figure 10-1.** *A site plan provides an overview of a job site, including the location of the point of beginning, utility easements and connections, and property lines.*

The contractor needs to check with the local utilities protection service to make sure all utilities in the area have been marked. Also, it is a good practice to mark the drawings with the location of the utility. The utility locator's mark on the ground is generally more accurate than an engineer's detail on a drawing.

## Benchmarks

A benchmark is a monument or mark placed by a surveyor that shows precise location and elevation. Benchmarks can range from brass disks cast in concrete (placed by the United States Geologic Survey) to a railroad spike driven into a telephone pole. The benchmarks are the point of reference for the entire job. The point of beginning is the place on the property line where the measurements begin. The point of beginning is located relative to a benchmark.

Surveyors often place corner stakes, known as hubs, at each corner of the property. **See Figure 10-2.** Traditionally, hubs were wooden stakes driven so that their tops were flush with the surface of the ground. Small nails driven into the top of each corner stake marked the exact corners of the property. Today, hubs are often a piece of rebar with a brightly colored plastic cap placed over the top end. Another method of marking a hub is to use a piece of pipe with a cork or lead plug.

Hubs form the control lines that will be used to layout the light pole bases, utility holes, conduit runs, and any other electrical work in that area. *Note:* Hubs mark only horizontal distance. They do not indicate elevation.

Drawings show the elevation of storm sewer castings and lines, water, gas, and sanitary sewer services, and any other underground utility relative to the benchmark or datum. A *datum* is a reference point to which other elevations, angles, or measurements are related. The datum may be the benchmark or it may be some other point a known distance and elevation from the benchmark.

Hubs

**Figure 10-2.** *A hub is used to mark a corner of the property.*

## Printreading

After the electrical section has been reviewed, the utility and grading plans sections should be carefully studied for any potential conflicts. The drawings use abbreviations for common terms. **See Figure 10-3.** For example, CB represents "catch basin," EP represents "edge of pavement," and TC represents "top of casting" or "top of curb." Other abbreviations should be shown in the drawing key.

The site plan typically shows contour lines. Contour lines are used when considering existing elevations and minimum fill requirements. For example, if a job has a minimum buried depth requirement of 4′, the trench must be deep enough that the buried depth is 4′ below the finished contour, not the existing contour.

---

### *T*ech Fact

The Construction Specifications Institute publishes the CSI MasterFormat™, which is a master list of the divisions and titles for organizing information about construction requirements and activities into a standard sequence. Each division is divided into subclassifications that provide greater detail.

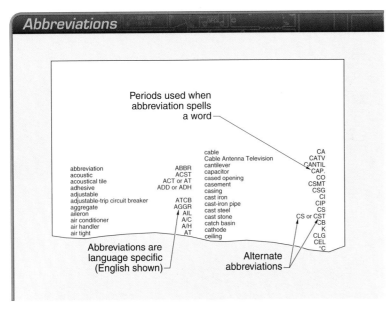

**Abbreviations**

Periods used when abbreviation spells a word

| | | | |
|---|---|---|---|
| abbreviation | ABBR | cable | CA |
| acoustic | ACST | Cable Antenna Television | CATV |
| acoustical tile | ACT or AT | cantilever | CANTIL |
| adhesive | ADD or ADH | capacitor | CAP. |
| adjustable | | cased opening | CO |
| adjustable-trip circuit breaker | ATCB | casement | CSMT |
| aggregate | AGGR | casing | CSG |
| aileron | AIL | cast iron | CI |
| air conditioner | A/C | cast-iron pipe | CIP |
| air handler | A/H | cast steel | CS |
| air tight | AT | cast stone | CS or CST |
| | | catch basin | CB |
| | | cathode | K |
| | | ceiling | CLG |
| | | | CEL |
| | | | °C |

Abbreviations are language specific (English shown)

Alternate abbreviations

*Figure 10-3. Abbreviations are frequently used on prints.*

When reading site plans, dimensions are given in feet and decimal feet. For example, an elevation or contour line marked on a drawing as 101.75 represents one hundred one and seventy-five hundredths feet, or 101′-9″.

## SURVEYING INSTRUMENTS

Surveying instruments are used to measure vertical and horizontal distance and direction to determine the precise location of an object relative to a benchmark. Surveying instruments range from inexpensive builder's levels to expensive global positioning systems that use satellites to figure elevation and coordinate differences. Other typical surveying instruments include engineer's levels, automatic levels, transits and theodolites, and EDM systems. Surveying instruments are not part of an electrician's standard tool list.

### Builder's and Engineer's Levels

A builder's level is an instrument consisting of a telescope with crosshairs and a spirit level. A builder's level has one plane of motion. An engineer's level is similar to a builder's level but has better optics and a micrometer-adjusting screw that allows more precise leveling.

A level is used to create a horizontal plane that is used as a datum for field excavation work. Instrument height (HI) is the elevation of a line of sight when using a surveying instrument. **See Figure 10-4.** A common use for a builder's or engineer's level is to find an unknown elevation from a known elevation.

Builder's and engineer's levels have a small round table on which the optics rotate, marked in degrees and fractions of a degree. **See Figure 10-5.** The horizontal circle and the vernier scale are used to measure horizontal angles. The scales extend for a full rotation.

A shot is a measurement made with a surveying instrument. The angle of the optics can be noted for two shots. The difference between the two angle measurements is the angle between the two shots.

It is important to note that once the instrument is leveled and a shot is taken of the known elevation, the reading of that shot is the datum in relation to the known elevation. This reading must be compared to the reading taken at the unknown point. In other words, the instrument datum is never at the benchmark elevation. For most applications, the value of the actual instrument height relative to the benchmark is unimportant. The difference between a known and unknown elevation is the important measurement.

### Automatic (Self-Leveling) Levels

Errors in instrument leveling are a common cause of mistakes in layout work. An automatic (self-leveling) level is a surveying instrument that employs a small pendulum and a compensator. As long as the tripod is fairly close to level, the instrument will level itself.

It is not common to find these instruments at a construction site. Self-leveling instruments do not perform well in areas that have a lot of vibration. Additionally, because of their sensitive instrumentation, special care must be taken when transporting and storing these valuable tools.

**Instrument Height**

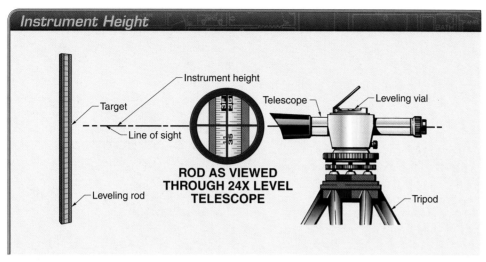

**ROD AS VIEWED THROUGH 24X LEVEL TELESCOPE**

**Figure 10-4.** *The line of sight is the level line extending from the horizontal crosshair at the center of the telescope barrel to the target.*

**Level Horizontal Circle**

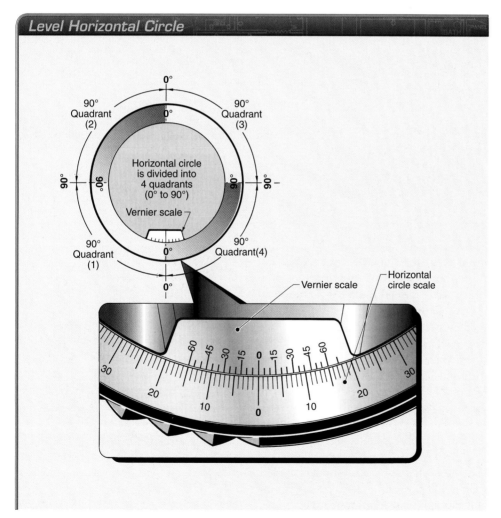

**Figure 10-5.** *Each quadrant of a horizontal circle scale is divided into degrees from 0° to 90°. The vernier scale is used to provide more accurate readings in minutes.*

## Transits and Theodolites

A transit is a surveying instrument with a telescope that rotates horizontally and vertically. In common usage, builder's levels and engineer's levels are often called transits. However, a true transit has both horizontal and vertical motion. This additional motion allows a transit to measure vertical angles, a capability not present in levels. Transits typically have a vertical arc. **See Figure 10-6.** The vertical arc typically extends to 45° above and 45° below instrument height.

A theodolite is a precision transit. Theodolites are extremely accurate and extremely expensive. An optical plummet is used to place the instrument directly over an object, such as a hub or surveyor's mark. Theodolites also employ a mirrored split-bubble that allows leveling accuracy far beyond a simple spirit vial. Typically, theodolites will measure within 10″ of arc. On a construction site, generally only the surveying crew will use these instruments.

## Electronic Distance Measurement and Global Positioning Systems

Electronic distance measurement (EDM) uses a surveying instrument that sends a beam of light or microwave energy to a prism mounted on or above the object being measured. **See Figure 10-7.** The time it takes for the light to be reflected off of the prism and return to the instrument is calculated. The instrument, using the speed of light as a known, computes the distance to the object. Relatively new theodolites incorporate EDM. These instruments can precisely measure angles and distances.

A global positioning system (GPS) is a surveying instrument that uses signals sent back and forth between satellites and the system. A GPS outputs precise rectangular coordinates and elevations and can compare one point to another. A GPS can be used with other types of surveying equipment.

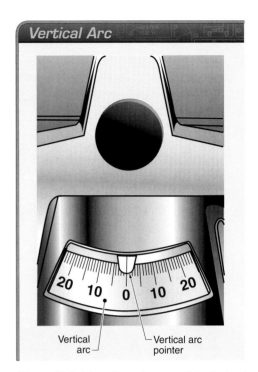

**Figure 10-6.** A transit may have a pointer instead of a vernier scale.

**Figure 10-7.** An EDM uses a surveying instrument that sends a beam of light or microwave energy to a prism mounted on or above the object being measured.

## Tripods and Leveling

Surveying instruments are generally mounted on a tripod. A tripod is a three-legged adjustable support for a surveying instrument. Tripods are not interchangeable and are usually designed for one instrument only. **See Figure 10-8.**

The location of the tripod and surveying instrument is critical. A location should be chosen that has a clear line of sight to the benchmark and to the objects being sighted. The accuracy of the survey will increase if the instrument remains in one position throughout the exercise. The tripod location should also be on solid ground as far away from heavy machinery and vibration as possible.

After finding a suitable location, the three legs of the tripod should be evenly spaced and the points driven into the soil. Each leg can then be adjusted to bring the apex of the tripod to a comfortable elevation. The small table at the apex of the tripod should be level.

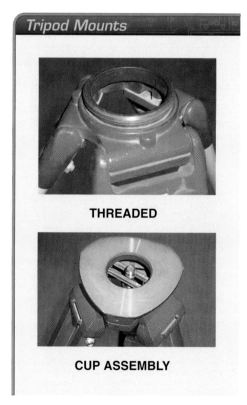

**Tripod Mounts**

**THREADED**

**CUP ASSEMBLY**

*Figure 10-8. Two types of tripod mounts are the threaded and the cup assembly.*

Normally every surveying instrument has a spirit level. This spirit level may be a single bubble (fish-eye) level as would be found on self-leveling instruments, or it may be an extremely accurate mirrored vial as would be found on theodolites. A spirit level works in the same manner as a torpedo level. However, a spirit level is much more accurate than a torpedo level. Mirrored levels employ the same vials as a spirit level, except a mirror is used to compare half the actual bubble with its mirrored image. Once the bubble and its reflection are aligned, the instrument is level.

**Adjusting Builder's Levels.** The telescope of a builder's level rotates on top of the circular base. To function properly, the telescope must be level in all positions over the base. Leveling a telescope is accomplished by adjusting the leveling screws. **See Figure 10-9.**

Although there should be firm contact between the leveling screws and base, overtightening the screws can damage the instrument. The leveling procedure becomes quick and simple with practice. When leveling a builder's level, opposite leveling screws must be turned equally, at the same time, and in opposite directions. The direction that the left thumb moves is the direction that the bubble moves. **See Figure 10-10.**

When a builder's level is used to lay out horizontal angles, it must be positioned directly over a specific point while it is being adjusted. However, a builder's level is not typically used for this application because it is easier to lay out horizontal angles with a transit level.

## Plumb Bobs

A plumb bob is a cone-shaped metal weight fastened to a string. **See Figure 10-11.** The force of gravity on this weight causes the string line to hang in a true vertical plane. Plumb bobs are suspended from the apex of most surveying instruments so the instruments may be set up exactly over a hub or benchmark. A plumb bob is also used when taking precise horizontal measurements with a tape.

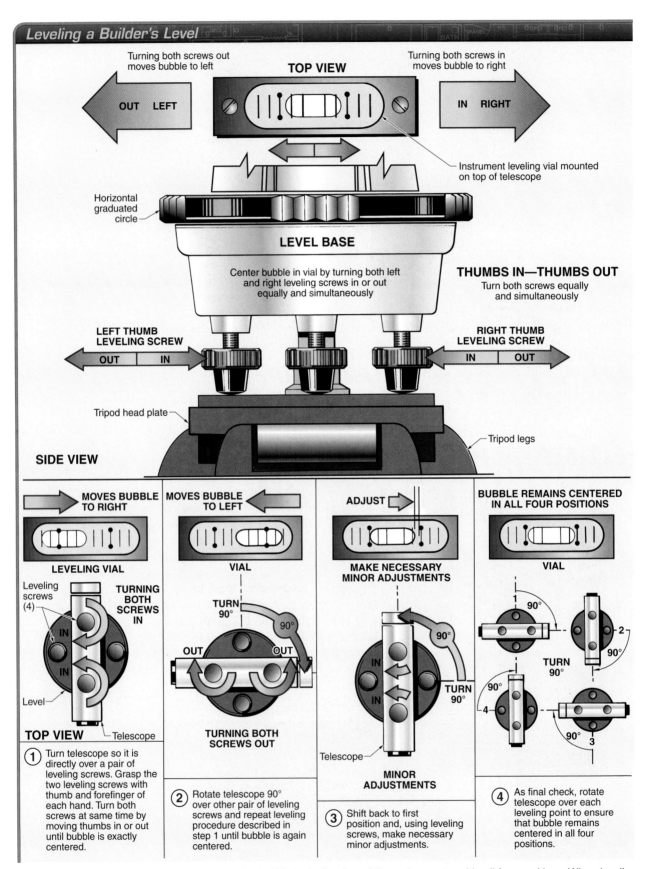

**Figure 10-9.** *A builder's level is level in which the bubble in the leveling vial remains centered in all four positions. When leveling a builder's level, the leveling screws must not be overtightened.*

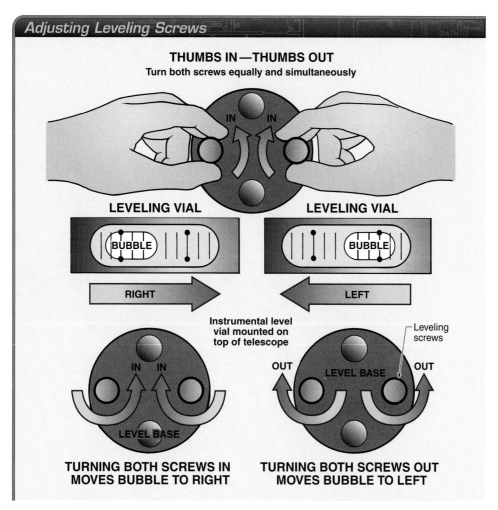

**Adjusting Leveling Screws**

**THUMBS IN—THUMBS OUT**
Turn both screws equally and simultaneously

IN    IN

LEVELING VIAL          LEVELING VIAL

BUBBLE                        BUBBLE

RIGHT                    LEFT

Instrumental level
vial mounted on
top of telescope

Leveling
screws

IN    IN              OUT    LEVEL BASE    OUT

LEVEL BASE

**TURNING BOTH SCREWS IN
MOVES BUBBLE TO RIGHT**

**TURNING BOTH SCREWS OUT
MOVES BUBBLE TO LEFT**

*Figure 10-10.* When leveling a builder's level, the direction the left thumb moves is the direction that the bubble moves.

**Plumb Bobs**

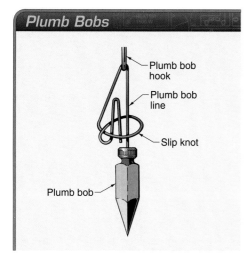

Plumb bob
hook

Plumb bob
line

Slip knot

Plumb bob

*Figure 10-11.* A plumb bob is used to locate a surveying instrument over a specified point on the ground. A slip knot permits the plumb bob to be raised or lowered as needed.

## Leveling Rods

A leveling rod is a measuring tool used with surveying instruments to lay out vertical elevations. **See Figure 10-12.** Leveling rods can be made of fiberglass or of wood. They are clearly marked so they can be seen through the sight of a surveying instrument at long distances. Leveling rods are placed firmly on top of an object and held plumb so a sighting can be made of the elevation of the object.

**Tech Fact**

A surveying instrument should always be allowed to acclimate to the surrounding temperature before use. The lenses can be gently cleaned with a lens tissue.

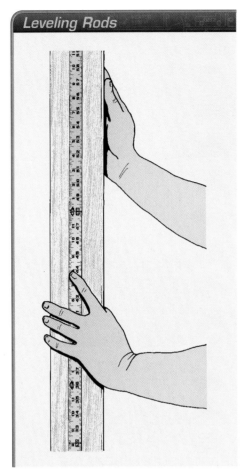

**Leveling Rods**

**Figure 10-12.** *Electricians often use a board and tape measure as a leveling rod.*

*Carlon*

*A trench may be considered a confined space. OSHA rules must be followed when working in a confined space.*

## EXCAVATION

An excavation is any cut, depression, or trench in the earth's surface formed manually or with earth-moving equipment. **See Figure 10-13.** A trench is a narrow excavation 15′ wide or less in which the depth is greater than the width. Electricians often work in excavations while laying underground conduit.

### Excavation Safety

As on all job sites, care must be taken when working around power equipment. Excavation equipment can cause serious injury to electricians working near it. Line trucks or digger derricks are often used to excavate light pole bases. Caution must be exercised when the augers of these trucks are in motion. If a glove or sleeve gets caught in a rotating auger, severe injury can occur.

**Confined Spaces.** A confined space is a space large enough and so configured that a worker can enter and perform assigned work, has limited or restricted means for entry and exit, and is not designed for prolonged occupancy. Confined spaces are subject to the accumulation of toxic or flammable contaminants, resulting in an oxygen-deficient atmosphere. Confined spaces include storage tanks, underground utility vaults, and pits and trenches more than 4′ deep. **See Figure 10-14.**

All employees entering a confined space must be instructed as to the nature of the hazards, necessary precautions, and protective and emergency equipment required. A safety harness with an attached lifeline must be worn when entering a confined space if a safe atmosphere cannot be ensured. Another worker, also wearing a safety harness and lifeline, must constantly observe the worker in the confined space.

*Tech Fact*

There must be at least one outside attendant whenever workers are inside a permit-required confined space. The attendant is responsible for periodically testing the atmospheric conditions and monitoring the workers inside.

**Excavations**

*Figure 10-13.* An excavation is a cut, depression, or trench in the earth's surface formed manually or with earth-moving equipment.

**Confined Spaces**

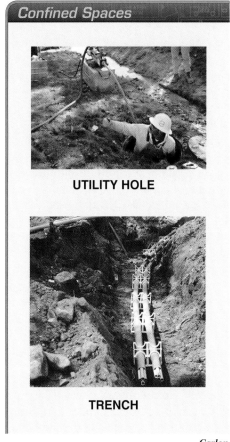

**UTILITY HOLE**

**TRENCH**

*Carlon*

*Figure 10-14.* Confined spaces include pits and trenches more than 4′ deep.

**Cave-ins.** A potential hazard when working in excavations is a cave-in. A cave-in occurs when the earth banks collapse. **See Figure 10-15.** OSHA regulations state that workers in excavations must be protected from cave-ins by adequate protective systems except when the excavation is made entirely in stable rock or when excavations are less than 5′ deep and inspection of the ground by a qualified person provides no indication of potential cave-ins.

Workers in excavations not made in stable rock or excavations more than 5′ deep must be protected by sloping, shoring, or an equivalent means. The method used to protect against collapsing earth banks depends on the soil type, excavation depth, water table level, type of foundation, and the space around the excavation. When working in or around excavations, the following safety rules must be observed:

• Excavated soil (spoil) and other materials and construction equipment must be at least 2′ away from the edge of the excavation.

• To provide a means of ingress and egress in case of an emergency, a ramp, runway, ladder, or stairway must be located within 25′ of workers if the excavation is 4′ or more in depth.

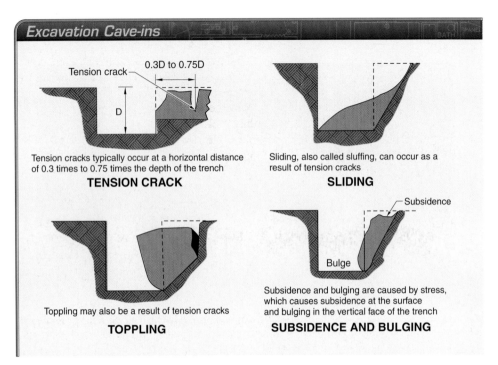

*Figure 10-15. Common causes of cave-ins are sliding and toppling, which cause tension cracks, and subsidence and bulging.*

- Vibration and increased lateral pressure on sides of excavations should be noted, such as soil vibrating loose when vehicles travel close to excavations.

- Adequate removal of engine exhaust from trenches must be provided if gasoline- or diesel-powered construction equipment is being used.

- Excavations containing water or in which water is accumulating should not be entered unless adequate precautions have been taken.

- Support systems such as shoring and bracing must be provided when the stability of adjoining buildings, walls, or other structures is endangered by excavation operations.

**Ingress and Egress.** Safe and immediate access to or exit from a trench must be available at all times to persons working in an excavation. Trenches that are 4′ deep or more must have a fixed means of exit. In addition, atmospheric testing must be conducted in excavations over 4′ deep where hazardous atmospheres can reasonably be expected to exist. Such locations might be near landfills, hazardous substance storage areas, or gas pipelines. Low oxygen levels and inflammable gases are two of the main hazardous atmospheres encountered. These circumstances require respiratory equipment and/or forced ventilation.

Ladders should be placed so that a worker will not have to travel more than 25′ in a lateral direction to climb out of a trench. Ladders must be stabilized and must extend at least 36″ above the ground surface. **See Figure 10-16.** Wooden ladders are preferred, particularly when electrical utilities are in the vicinity.

Some construction jobs may require ramps or temporary stairs as a means for workers to move around on the job and for materials to be transported. Ramps and runways may also be constructed for wheelbarrows and power buggies used to transport materials.

**Water Accumulation.** The accumulation of water can cause problems at an excavation. Excavation in areas subject to runoff from heavy rain or in close proximity to natural water must have a special support or shield system because saturated soil can cave in.

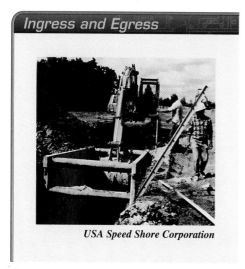

**Figure 10-16.** *Ladders must be provided for ingress and egress in trenches over 4′ deep.*

Water removal equipment must be on hand to control water accumulation. Ditches and dikes can be constructed to prevent water from entering the excavation site. Steps should be taken to provide adequate drainage of the areas adjacent to the excavation site.

A good rule of thumb is to open the trench only enough for one day of work. If possible, the run started that day should be placed, backfilled, and compacted by the end of that day. This will minimize the effects of an overnight rain.

## Sloping and Benching

Sloping is the process of cutting back trench walls to an angle that reduces the chance of collapse into the work area. Benching is sloping that cuts back trench walls into a step pattern. **See Figure 10-17.** Vertical walls produced by benching cannot exceed 3½′ in height. Sloping is typically used when there is room around the construction area to slope the sides of an excavation. Sloping cannot be used if buildings or streets are immediately adjacent to the excavation.

**Tech Fact**

According to OSHA, trenches should be inspected before construction begins and as needed during the day.

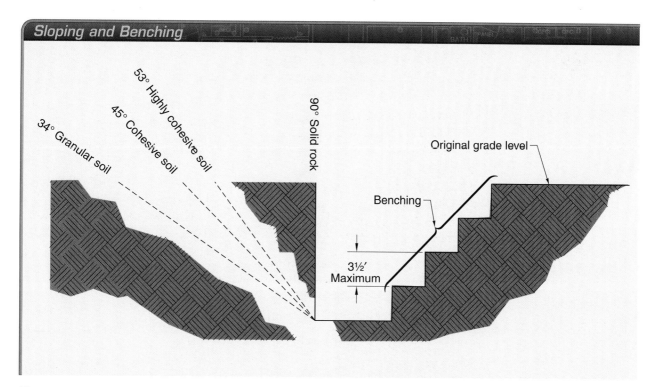

**Figure 10-17.** *Sloping and benching are used to prevent cave-ins in excavations and trenches.*

The amount of slope required is determined by soil conditions and Occupational Safety and Health Administration (OSHA) standards. OSHA regulations specify sloping at a 34°, 45°, or 53° angle, depending on the soil type. Sloping can be expensive because it commonly requires the acquisition of expensive rights of way in addition to excess excavation, backfilling, and compaction costs. Shoring is required at job sites where sloping cannot be used.

## Shoring

Shoring is the use of wood or metal members to temporarily support soil, formwork, or construction materials. The shoring method used is determined by the requirements of the job site in compliance with the regulations of the authority having jurisdiction. Shoring includes vertical shoring, walers, steel soldier piles and wood lagging, and wood shoring. **See Figure 10-18.**

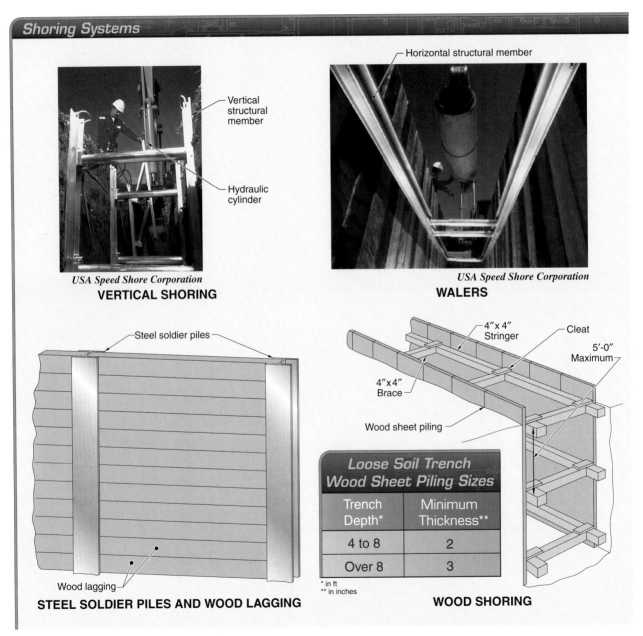

### Shoring Systems

**VERTICAL SHORING**
USA Speed Shore Corporation

Vertical structural member

Hydraulic cylinder

**WALERS**
USA Speed Shore Corporation

Horizontal structural member

**STEEL SOLDIER PILES AND WOOD LAGGING**

Steel soldier piles

Wood lagging

**WOOD SHORING**

4" x 4" Stringer

Cleat

5'-0" Maximum

4" x 4" Brace

Wood sheet piling

| Loose Soil Trench Wood Sheet Piling Sizes | |
| Trench Depth* | Minimum Thickness** |
| --- | --- |
| 4 to 8 | 2 |
| Over 8 | 3 |

* in ft
** in inches

**Figure 10-18.** Shoring systems provide temporary support to prevent soil or construction materials from caving into an excavation.

Vertical shoring is shoring that uses opposing vertical structural members with cross bracing of screw jacks or hydraulic or pneumatic cylinders. Vertical shoring is available as pre-engineered aluminum components that are installed and removed from the top of the trench. Vertical shoring is commonly installed and removed by one worker and used in soil with good cohesion. Shields (retaining components) made from fiberglass, wood, or metal and placed between the vertical structural members and the soil can be used to provide additional protection.

A waler is a horizontal support member used to hold trench sheet piling. Walers are designed to support a variety of retaining members and are used in unstable soil. Waler shoring systems are installed and removed from aboveground by hand or with excavation equipment. Walers are used to provide support in large expanses.

Large, deep excavations often require the use of steel soldier piles and wood lagging. A soldier pile is a vertical steel H-beam that is driven into the ground. Lagging is planking used to retain earth on the side of a trench or excavation. Wood lagging is placed between the steel soldier piles. Wood shoring is shoring that uses wood components for stringers, braces, and piling.

## Shielding

A shield is a portable protective device capable of withstanding forces from a cave-in. Shielding is used for deep and/or wide excavations and commonly uses trench boxes. A trench box is a type of shield that uses two plates held apart by spacers to shore the sides of a trench. Trench boxes are made from steel, concrete, or wood and are moved along the trench as work progresses. Trench boxes allow for excavation and backfilling to occur while work is being done within the shield. **See Figure 10-19.**

Trench boxes can be used in stable or unstable soil if proper excavation techniques are used. In stable soil, the trench is excavated to the proper grade, slightly wider than the width of the trench box. The trench box is placed in the trench and excavation is continued in front of the shield. The trench box is pulled forward and the trench is backfilled as work progresses.

In unstable soil, the trench is excavated until the soil does not crumble into the desired trench width. The trench box is placed in the excavated area and each end is alternately pushed down until reaching the proper grade.

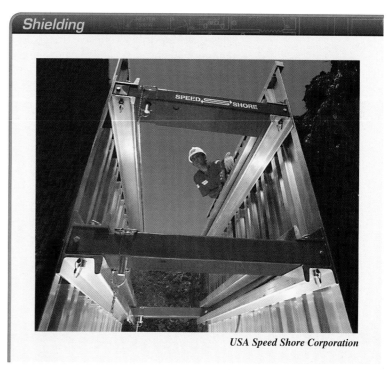

*USA Speed Shore Corporation*

**Figure 10-19.** *Shields provide cave-in protection for workers and are moved by excavation equipment as work progresses.*

## UNDERGROUND INSTALLATION PROCEDURES

After an excavation is complete, the installation of the conduit begins. The job specifications give the details on installation. Some installations are made with PVC conduit. Others are made with steel conduit. Some installations are concrete encased. Others are covered with gravel or other stone.

> **Tech Tip**
>
> Steel elbows are sometimes used in the middle of PVC runs. Pulling wire on high-tension pulls can cut through PVC elbows.

## Concrete

In many cases, a PVC conduit run will be encased in concrete to protect the plastic conduit from damage. Electricians also use concrete when placing light pole bases and transformer pads. Concrete is a mixture of Portland cement, sand, gravel (aggregate), water, and admixtures.

**Concrete Components.** Portland cement is a ground and calcined (heated) mixture of limestone, shells, cement rock, silica sand, clay, shale, iron ore, gypsum, and clinker. **See Figure 10-20.** When Portland cement is mixed with water it becomes gel-like in consistency and very cohesive. It acts as a binding agent in concrete. Portland cement gives off heat during the curing process and has the unique ability to set up under water. Portland cement mixed with sand and lime is known as mortar. When mixed with pea gravel, it is known as structural grout, and when mixed with aggregate, Portland cement becomes concrete.

Aggregates and sand make up 75% to 80% of concrete. Aggregates must be clean, free of silt, salt, or clay. They also must be structurally sound, stable, tough, hard, and durable. The shape of aggregate is also critical. The stone cannot be flat, elongated, or exceedingly round.

Admixtures are materials that are added to concrete to improve its characteristics. The three basic admixtures are pumice or fly ash, calcium chloride, and air-entraining agents. Pumice or fly ash is added to concrete to reduce the heat of hydration and also as a binder for holding the mixture together. Calcium chloride is added to concrete to accelerate the set-up time and is often used in the winter months. Air-entraining agents, which are natural or synthetic soaps, are added to concrete to form air bubbles to act as stress relievers during freeze-thaw cycles.

**Concrete Types.** There are five main types of Portland cement. Type I is general-purpose cement. Type II is modified general-purpose cement that is able to withstand some chemicals. Type III is high early-strength cement. Type IV is low-heat cement that is used in large mass placings where the heat of hydration would impose undesirable stress. Type V is sulphate-resistive cement that is used where the cement will be in contact with alkaline water.

### Tech Fact

Portland cement can be stored indefinitely in dry conditions without losing any of its properties.

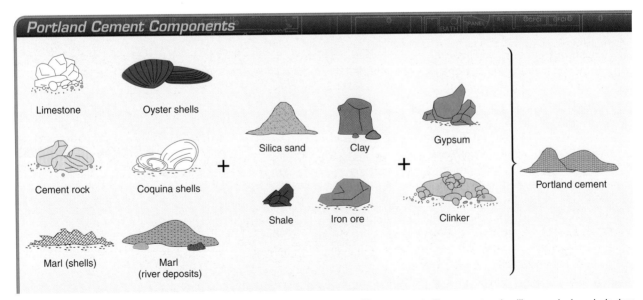

**Portland Cement Components**

Limestone — Oyster shells — Silica sand — Clay — Gypsum

Cement rock — Coquina shells — + — Shale — Iron ore — + — Clinker — Portland cement

Marl (shells) — Marl (river deposits)

**Figure 10-20.** Portland cement is a ground and calcined (heated) mixture of limestone, shells, cement rock, silica sand, clay, shale, iron ore, gypsum, and clinker.

**Concrete Testing.** Water must be added to cement to start the chemical reaction that hardens the cement into concrete. As a general rule, the more water that is mixed with cement the weaker the mixture becomes. A slump test measures the consistency of fresh concrete. **See Figure 10-21.**

A test laboratory may be present to test the concrete slump. In a slump test, a cone-shaped mold is filled with wet concrete and tested according to ASTM C143, *Standard Test Method for Slump of Hydraulic Cement Concrete.* A slump test measures the consistency of the concrete by measuring how much the concrete slumps, or falls, when the cone is removed. A stiff mixture is indicated by 0″ to 2″ of slump. A low/medium mixture is indicated by 2″ to 4″ of slump. A wet mixture is indicated by 4″ to 6″ of slump. A flowing mixture is indicated by over 6″ of slump.

Concrete used to protect conduits may be exempt from the slump requirement since the concrete is not structural. Concrete used for light pole bases and transformer pads should be tested if provisions for testing are available.

## BACKFILLING

After a trench has been opened, the conduit installed, and the concrete placed and allowed to cure, the trench must be filled again. Backfilling is the replacement of soil from an excavation. The job specifications should have the necessary details on the type of backfill material and the required compaction.

### Backfilling Materials

The lowest level of a trench is often filled with stone or gravel. Graded stone is typically used as backfill material. Generally, the smaller the grade number, the larger the stone. For example, number 2 stone has a diameter of approximately 1½″ to 2″, while number 57 stone has a diameter of ¾″. Other popular backfill materials include 304 and 46D stone. Both of these types of stone

are self-compacting. They are composed of limestone aggregate with sizes from ¾″ down to the size of dust particles.

After the bottom of the trench has been filled with stone, the remainder of the trench is usually filled with soil. The soil used for backfill should be free of wood scraps and any other type of waste material. The job specifications may limit the amount of rock or other debris near the top of the trench.

Finally, the entire excavation should be finished and trimmed. The finish will vary depending on the job specifications. The conduit run may need to pass under a building wall, road, or parking lot. In a situation like this, the finish will be completed by one of the other construction trades. The conduit run may need to be finished with topsoil, sod, and landscaping.

### Compaction

Compaction is another important element to consider when backfilling. Compaction is the process of applying downward pressure to loose soil or stone to increase its density and load-bearing capacity through consolidation and removal of voids.

*Slump Test*

***Figure 10-21.*** *A slump test is used to measure the consistency of concrete.*

During excavating and trenching, the air that infiltrates the soil causes air voids and an increase in soil volume. Air voids cause soil weakness and an inability of the soil to carry heavy loads. Air voids in uncompacted soil can also cause undesirable settlement, resulting in sinking of the soil and causing cracks or complete failure of the trench.

Compacted soil reduces water penetration and related problems. Water penetration swells the soil. In addition, water expands when frozen in the soil, causing heaving and cracking of walls and floor slabs.

Many specifications call for a 95% to 100% compaction. A testing lab on site will ensure that these specifications are met. **See Figure 10-22.** Some stone is impossible to compact. For example, number 57 limestone will not compact to 90%. The type and shape of the stone cause it to move around within the trench but not compact. Other types of stone will offer 100% compaction quite easily. Any of the self-compacting stone will produce 100% compaction with just two passes of a compactor.

*Carlon*

*A trench can be very narrow when conduit is to be placed only a short depth below ground.*

**Compaction Methods.** Common compaction methods include impact force, vibration, and static force. **See Figure 10-23.** Impact force compaction is compaction using a machine that alternately strikes and leaves the ground. Cohesive (fine-grained) soils are best compacted by impact force. Cohesive soils do not settle under vibration due to natural binding between small soil particles. Impact force compaction on cohesive soils produces a shearing effect that squeezes air pockets and excess water to the surface and moves soil particles closer together.

Vibration compaction is compaction using a machine to apply high-frequency vibration to the soil to increase soil density. Granular (coarse-grained) soils are best compacted by vibration. Vibration motion allows soil particles to twist and turn into voids, which limits their movement. A combination of impact force compaction and vibration compaction may be needed to achieve compaction of granular soils.

Static force compaction is compaction using a heavy machine that squeezes soil particles together without vibratory motion to increase soil density. Static force compaction is rarely used because of the development of small efficient soil-compaction equipment.

*Soil Compaction Testing*

**Figure 10-22.** *Moisture/density gauges are used to determine the density of soil after the soil compaction process is complete.*

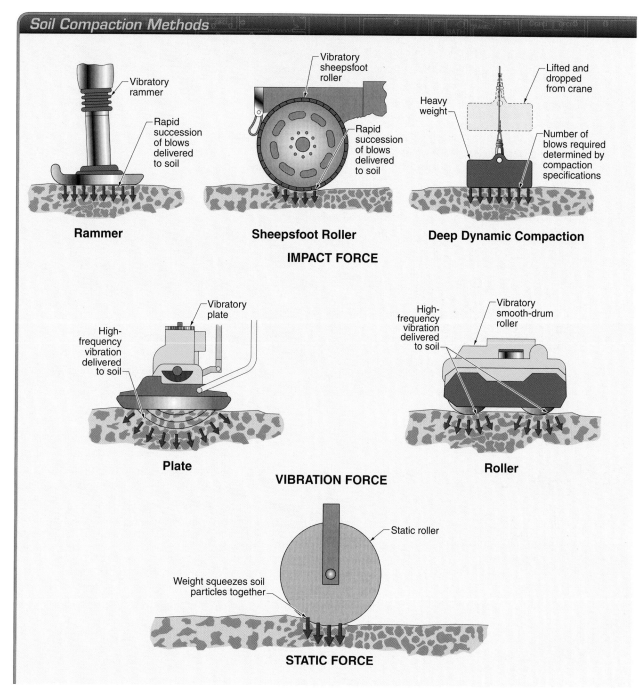

**Figure 10-23.** *Soil compaction is achieved using equipment that produces an impact, vibration, or static force.*

## SUMMARY

- A job specification typically comes with a site plan, mechanical drawings, and electrical drawings.

- A benchmark is a monument or mark placed by a surveyor that shows precise location and elevation.

- Hubs are stakes that form control lines that are used to lay out light pole bases, utility holes, conduit runs, and any other electrical work in the area.

- A builder's level is an instrument consisting of a telescope with crosshairs and a spirit level. A builder's level has one plane of motion.

- A transit is a surveying instrument with a telescope that rotates vertically and horizontally. A true transit has both vertical and horizontal motion.

- A tripod is a three-legged adjustable support for a surveying instrument. Tripods are not interchangeable and are usually designed for one instrument only.

- A potential hazard when working in excavations is a cave-in, where the earth banks collapse.

- All employees entering a confined space must be instructed as to the nature of the hazards, necessary precautions, and protective and emergency equipment required.

- Ladders should be placed so that a worker will not have to travel more than 25′ in a lateral direction to climb out of a trench. Ladders must be stabilized and must extend at least 36″ above the ground surface.

- Sloping is the process of cutting back trench walls to an angle that eliminates the chance of collapse into the work area.

- Shoring is the use of wood or metal members to temporarily support soil, formwork, or construction materials in a trench.

- Concrete is a mixture of Portland cement, sand, gravel (aggregate), water, and admixtures.

- Compaction is the process of applying downward pressure to loose soil or stone to increase its density and load-bearing capacity through consolidation and removal of voids.

- Common compaction methods include impact force, vibration, and static force.

# Appendix
# Table of Contents

## A Appendix A

Math Review ................................................................ 177

## B Appendix B

Trigonometry for Conduit Bending ................................ 183

Finding an Unknown Radius ........................................ 192

## C Appendix C: Conduit Bending Supplements

Trigonometry Table .................................................... 197

Offset Multiplier (Cosecant Table) ............................. 198

Bending Quick Guide ................................................. 199

  Take-Up .................................................................. 199

  Hand Bending Gain ................................................ 199

  Distance Multiplier and Shrink Constant ............... 199

  Three-Bend Saddle Multiplier ............................... 199

Conduit Dimensions .................................................. 200

  GRC Rigid .............................................................. 200

  EMT ....................................................................... 200

  PVC Schedule 20 ................................................... 200

  PVC Schedule 40 ................................................... 200

  PVC Schedule 80 ................................................... 200

# Appendix
# Table of Contents

*Aluminum Rigid* ........................................................................................*200*

*Aluminum EMT* ........................................................................................*200*

*Conduit Layout Guide* ..............................................................................*201*

*Minimum Center-to-Center Measurement* ...............................................*201*

*Conduit Bushing Diameters* ....................................................................*201*

*Conduit Locknut Diameters* .....................................................................*201*

*Expansion Characteristics of Steel Conduit and Tubing* ........................*201*

*Decimal Feet and Inches Conversions* ...................................................*202*

*Fractional and Decimal Conversions* ......................................................*202*

*Thread Protector Cap Colors* ..................................................................*202*

*Conduit Bending and Fabrication Products* ............................................*203*

# Appendix A: Math Review

## ARITHMETIC REVIEW

Measurements used in conduit bending are either provided on a drawing or taken in the field. These measurements are usually in feet, inches, and fractions of an inch. The electrician must be able to locate the measurements on a rule or tape measure. In some cases, the measurements must be added or subtracted to achieve the desired results.

### Reading Rule and Tape Measure Markings

There are different types of markings on different measuring tools. However, all rules and tape measures have several markings in common.

The most prominent markings on a tape measure are the large numbers at every inch, located next to the long lines. **See Figure A-1.** These long lines usually extend all the way across the tape. The numbers designate the number of inches from the end. Every 12″ there is another number, which represents the number of feet from the end. In addition, there are smaller markings between the inch lines. These markings represent fractions of an inch.

Each inch is divided into 16 equal divisions. Each division represents ¹⁄₁₆″. Between the inch lines, the longest marks represent halves of an inch and the next longest marks after that represent quarters of an inch. After that, the marks continue to get smaller for eighths and sixteenths of an inch.

Some measuring tools may have markings representing other divisions. For example, some tools have markings every ¹⁄₃₂″ for the first foot and every ¹⁄₁₆″ after that. Some tools have additional mark-

ings at other intervals for carpentry applications, such as every 16″ for locating studs. Others have standard inch and foot measurements along one edge and metric measurements along the other edge. These extra markings can usually be ignored in conduit bending applications.

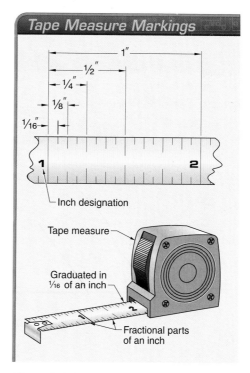

**Figure A-1.** A tape measure or rule has markings to indicate the inches and fractions of an inch.

### Using Fractions

Fractions of an inch are expressed with a numerator and a denominator. The numerator and denominator are separated by a horizontal or inclined fraction bar. The denominator is the lower number of the fraction. For example, in the fraction ⁵⁄₁₆″,

16 is the denominator. The denominator indicates the number of equal divisions of the inch. A greater number of divisions indicates a greater degree of precision.

The numerator is the upper number in the fraction. The numerator shows the number of actual divisions that make up a given length of measurement. For example, $5/16''$ indicates an inch that has been divided into 16 equal parts (the denominator), of which the actual length of measurement is equal to 5 of those parts (the numerator).

The distance between the smallest marks on a tape measure represents a length of $1/16''$. For example, the distance from the 12″ mark to the next small mark is $1/16''$. **See Figure A-2.** A distance of two marks is $2/16''$, representing 2 parts of the 16 equal parts of an inch.

The numerator of a fraction is normally an odd number. If the numerator is an even number, these fractional values are converted to simpler equivalent fractions. For example, $2/16''$ is equal to $1/8''$. Some values can be further simplified. The fraction $8/16''$ is equal to $4/8''$, $2/4''$, and $1/2''$. This simplification is used when adding and subtracting fractions of an inch.

## Using Mixed Numbers

A mixed number is a combination of a whole number and a fraction. For example, a distance measurement of $12\,5/16''$ is a mixed number that is a combination of the number 12, representing 12″, and the fraction $5/16$, representing $5/16''$. The number and the fraction are added to represent the total distance. **See Figure A-3.**

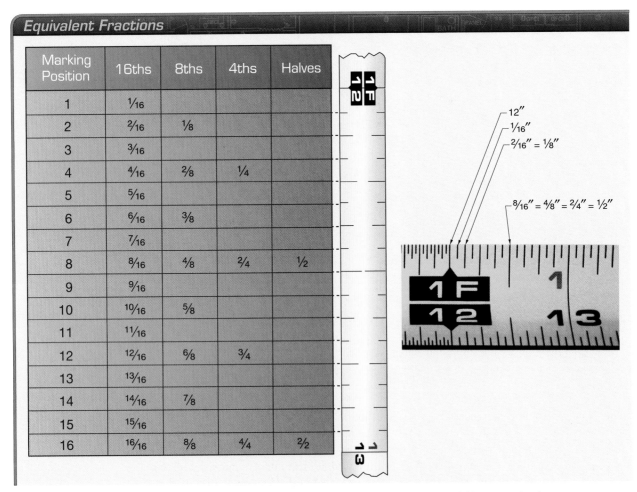

**Figure A-2.** *Many fractions are equivalent to other fractions. These equivalent fractions should be memorized.*

An improper fraction, such as $^{15}/_{12}''$, has a numerator larger than its denominator. An improper fraction is normally reduced to a mixed number. The numerator is divided by the denominator, and the remainder is kept as a fraction. For example, 15 is divisible by 12 only 1 time with a remainder of 3. Therefore, $^{15}/_{12}''$ is equal to $1^3/_{12}''$, or 1'-3''.

A measurement in feet and inches is treated in a similar manner to a measurement in inches and fractions of an inch. An inch represents $^1/_{12}$ of a foot. For example, a measurement of 5'' represents a distance of $^5/_{12}'$. A distance of 15'' represents a distance of $^{15}/_{12}'$, or $1^3/_{12}'$, or 1'-3''. Similarly, a measurement of 2'-7$^1/_4$'' represents a measurement of 2' plus 7'' plus $^1/_4$''.

A measurement in feet can be converted to its equivalent measurement in inches by multiplying it by 12. For example, a measurement of 1' is equal to 12'' (1 × 12 = 12), and a measurement of 4' is equal to 48'' (4 × 12 = 48).

There are many situations where a measurement in feet and inches must be converted to inches. To accomplish this, the number of feet in the measurement are multiplied by 12 and added to the inch part of the measurement. For example, a measurement of 1'-1'' is equal to 13'' (1 × 12 + 1 = 13), and 4'-3'' is equal to 51'' (4 × 12 + 3 = 51).

## Adding Fractions and Mixed Numbers

In the field, dimensions on prints as well as other foot and inch measurements will need to be added together. Calculations involving feet are simple. Calculations involving inches are more complex because inches are based on 12 equal divisions of a foot, and fractions of an inch are based on 16 equal divisions.

Since the foot, inch, and fraction systems are based on different divisions, foot, inch, and fraction calculations must be performed separately and then combined. In order to add fractions, both fractions must have the same denominator, called a common denominator.

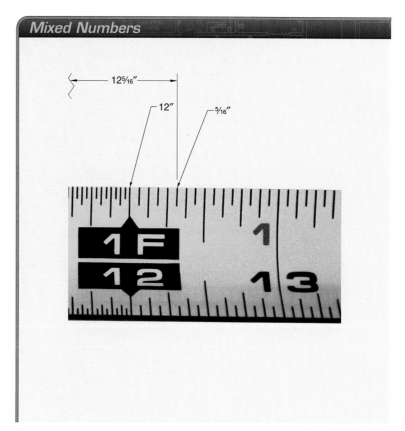

**Figure A-3.** *A mixed number is a combination of a whole number and a fraction.*

To add two fractions, the numerator and denominator of the fraction with the smaller denominator are multiplied by a factor so that the denominator equals the larger number denominator. Once the common denominator is determined, the numerators are added. For example, when adding $^1/_4''$ and $^5/_8''$, both the numerator and denominator of $^1/_4''$ are multiplied by a factor of 2 to change the fraction to $^2/_8''$. **See Figure A-4.** (It may be necessary to multiply the numerators and denominators of both fractions by a factor in order for the denominators to be equal.)

With a common denominator of 8, $^2/_8''$ and $^5/_8''$ are added for a total $^7/_8''$ ($^2/_8 + ^5/_8 = ^7/_8$). When adding $^1/_2''$ and $^5/_{16}''$, both the numerator and denominator of $^1/_2$ are multiplied by a factor of 8 to change the fraction to $^8/_{16}$. With a common denominator of 16, $^8/_{16}''$ and $^5/_{16}''$ are added for a total of $^{13}/_{16}''$ ($^8/_{16} + ^5/_{16} = ^{13}/_{16}$).

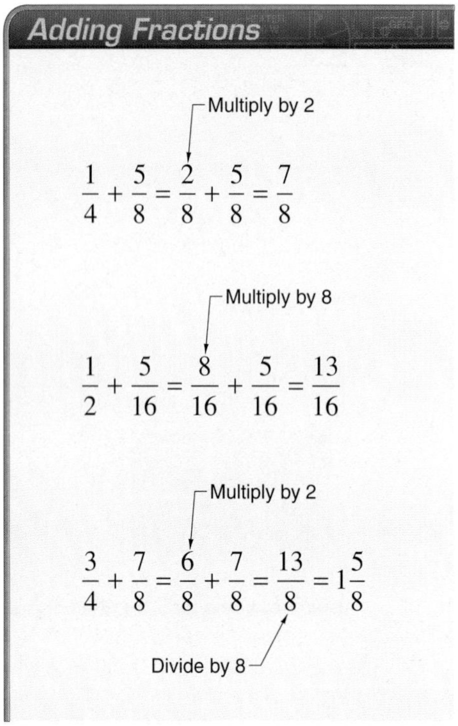


***Figure A-4.*** *Fractions can be added by using a common denominator.*

When the sum of two fractions is more than 1, the sum must be converted to a mixed number. For example, when adding $\frac{3}{4}''$ and $\frac{7}{8}''$, both the numerator and denominator of $\frac{3}{4}''$ are multiplied by 2 to change the fraction to $\frac{6}{8}''$. With a common denominator of 8, $\frac{6}{8}''$ and $\frac{7}{8}''$ are added for a total $\frac{13}{8}''$ ($\frac{6}{8} + \frac{7}{8} = \frac{13}{8}$). This fraction must be converted to a mixed number. The numerator is divided by the denominator and the remainder is kept as a fraction. For example, 13 is divisible by 8 only 1 time with a remainder of 5. Therefore, $\frac{13}{8}''$ is equal to $1\frac{5}{8}''$.

Length measurements are often given as mixed numbers. For example, a length is measured at $14\frac{1}{4}''$, and another length is measured as $35\frac{1}{8}''$. **See Figure A-5.** The whole inch part and the fraction part of the mixed numbers must be added separately, and the common denominator for the fraction part must be found first. In this case, the measurement of $14\frac{1}{4}''$ is equal to $14\frac{2}{8}''$. The sum of these two lengths is calculated as follows:

$$\begin{array}{r} 14\frac{2}{8}'' \\ +\ 35\frac{1}{8}'' \\ \hline 49\frac{3}{8}'' \end{array}$$

In some cases, the sum of the fraction part is more than 1. For example, a length is measured at $35\frac{7}{8}''$ and another length is measured as $14\frac{1}{4}''$. Again, the inch part and the fraction part of the mixed numbers are added separately, and the common denominator for the fraction part is found first. In this case, the measurement of $14\frac{1}{4}''$ is equal to $14\frac{2}{8}''$. The sum of these two lengths is calculated as follows:

$$\begin{array}{r} 35\frac{7}{8}'' \\ +\ 14\frac{2}{8}'' \\ \hline 49\frac{9}{8}'' \end{array}$$

The fraction part is an improper fraction that must be converted to a mixed number. The fraction $\frac{9}{8}$ is equivalent to $1\frac{1}{8}$. Therefore, the sum of $35\frac{7}{8}''$ and $14\frac{1}{4}''$ is $50\frac{1}{8}''$.

**Tech Fact**

Fractions need to be converted to decimal numbers before being added on a calculator.

## Subtracting Fractions and Mixed Numbers

Foot and inch measurements on job sites or dimensions on prints may need to be subtracted from each other. Subtracting fractions is very similar to adding fractions. The fractions must be converted to fractions with a common denominator. The numerator and denominator of the fraction with the smaller denominator is multiplied by a factor so that the denominator equals the larger denominator. (It may be necessary to multiply the numerators and denominators of both fractions by a factor in order for the denominators to be equal.)

Once the common denominator is determined, the numerators are subtracted. For example, when subtracting $\frac{1}{4}''$ from $\frac{5}{8}''$, both the numerator and denominator of $\frac{1}{4}''$ are multiplied by a factor of 2 to change the fraction to $\frac{2}{8}''$. **See Figure A-6.** With a common denominator of 8, $\frac{2}{8}''$ is subtracted from $\frac{5}{8}''$ for a difference of $\frac{3}{8}''$ ($\frac{5}{8} - \frac{2}{8} = \frac{3}{8}$).

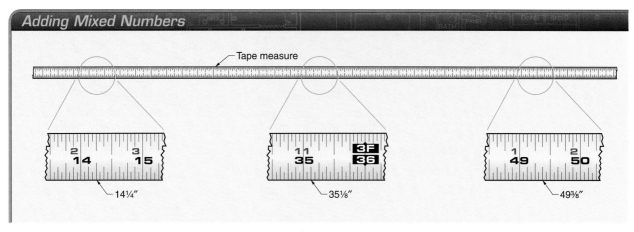

**Figure A-5.** *Adding mixed numbers to determine a total distance is often necessary on a job site.*

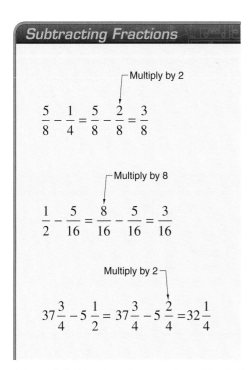

**Figure A-6.** *Mixed numbers can be subtracted by using a common denominator.*

When subtracting ⁵⁄₁₆″ from ½″, both the numerator and denominator of ½ are multiplied by a factor of 8 to change the fraction to ⁸⁄₁₆. With a common denominator of 16, ⁵⁄₁₆″ is subtracted from ⁸⁄₁₆″ for a difference of ³⁄₁₆″ (⁸⁄₁₆ – ⁵⁄₁₆ = ³⁄₁₆).

There are situations where mixed numbers must be subtracted. Just like with addition, the inch part and the fraction part are subtracted separately. The common

denominator for the fraction part must be found first. For example, a distance is measured at 37¾″. A length of 5½″ must be subtracted from the measurement. The numerator and denominator of the ½ must be multiplied by a factor of 2 to give ²⁄₄. The calculation is completed as follows:

$$
\begin{array}{r}
37\tfrac{3}{4}'' \\
-\ 5\tfrac{2}{4}'' \\
\hline
32\tfrac{1}{4}''
\end{array}
$$

## Converting between Fractions and Decimals

It is very useful to be able to convert between fractions of an inch and their decimal equivalents. Many calculations involving fractions are done on calculators. The fractions must be converted to decimal numbers so that they can be entered into the calculator. After the calculations are completed, the decimal number needs to be converted back to a fraction in order to use a tape measure.

The simplest method for converting between decimals and fractions is to use a conversion table. **See Figure A-7.** To convert from a fraction to a decimal, the fraction is located in the table and the equivalent decimal is found by following the line across in the table. To convert from a decimal to a fraction, the decimal number is found and the line followed back to its equivalent fraction.

| Fraction and Decimal Conversion Tables | |
|---|---|
| Fraction | Decimal |
| 1/16 | 0.0625 |
| 2/16, 1/8 | 0.125 |
| 3/16 | 0.1875 |
| 4/16, 2/8, 1/4 | 0.25 |
| 5/16 | 0.3125 |
| 6/16, 3/8 | 0.375 |
| 7/16 | 0.4375 |
| 8/16, 4/8, 3/4, 1/2 | 0.5 |
| 9/16 | 0.5625 |
| 10/16, 5/8 | 0.625 |
| 11/16 | 0.6875 |
| 12/16, 6/8, 3/4 | 0.75 |
| 13/16 | 0.8125 |
| 14/16, 7/8 | 0.875 |
| 15/16 | 0.9375 |
| 16/16 | 1 |

*Figure A-7. It is often necessary to convert a number from a fraction to a decimal or from a decimal to a fraction. These conversions should be memorized.*

For example, from the table the fraction 5/16 is equal to 0.3125. Similarly, if the result of a calculation is 0.625, then the equivalent fraction is 5/8. If the decimal number is not in the table, the value in the table that is nearest to a decimal should be selected. For example, if the result of a calculation is 0.6, the nearest decimal in the table is 0.625 and the equivalent fraction is 5/8.

If a calculator is available, any fraction can be converted to a decimal by dividing the numbers. For example, the fraction 3/4 can be converted to a decimal by dividing the numerator, 3, by the denominator, 4, resulting in 0.75. Similarly, the fraction 5/16 can be converted to a decimal by dividing the 5 by the 16, resulting in 0.3125.

A calculator can also be used to convert a decimal number to a fraction. Simply multiply the decimal number by 16. This gives the number of 16ths in the fraction. For example, the decimal number 0.62 multiplied by 16 gives 9.92, or approximately 10. Therefore, the decimal number 0.62 is approximately equal to 10/16, or 5/8. **See Figure A-8.**

For mixed numbers, care must be taken to multiply only the decimal part by 16. Multiplying the entire mixed number by 16 gives the wrong results. Only the part to the right of the decimal point is multiplied by 16. If a calculation result of 20.80″ is multiplied by 16, the result is 332.8, which has no meaning. If the part to the right of the decimal point, 0.80, is multiplied by 16, the result is 12.8, or about 13. This means that 20.80″ is approximately equal to 20^13/16″.

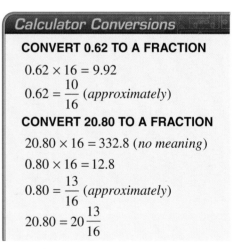

**Calculator Conversions**

**CONVERT 0.62 TO A FRACTION**

$0.62 \times 16 = 9.92$

$0.62 = \dfrac{10}{16}$ (*approximately*)

**CONVERT 20.80 TO A FRACTION**

$20.80 \times 16 = 332.8$ (*no meaning*)

$0.80 \times 16 = 12.8$

$0.80 = \dfrac{13}{16}$ (*approximately*)

$20.80 = 20\dfrac{13}{16}$

*Figure A-8. A calculator can be used to convert decimals to fractions.*

# Appendix B:
# Trigonometry for Conduit Bending

## TRIGONOMETRY

An understanding of basic trigonometry is useful in conduit bending. The study of trigonometry includes the study of angles and right triangles and the relationships between them. The term "trigonometry" is often shortened to "trig" in common usage.

## Angles

An angle is a measure of the rotation between two lines. **See Figure B-1.** Angles are usually measured in degrees and are designated with a degree symbol (°). The angle increases as the rotation between the lines increases. There are several special angles that are commonly used in conduit bending. Those angles are 10°, commonly used for box offsets; 15°, 22½°, 30°, 45°, and occasionally 60°, used for bending offsets and saddles; and 90°, used for fabricating 90° bends and stub-ups.

A right angle is a 90° angle. A very simple rule about angles and triangles is that the sum of all the angles is 180°. This means that if one angle in a triangle is 90° and another angle is 30°, then the third angle is 60°. This is true because 90 + 30 + 60 = 180.

## Right Triangles

A fundamental part of trigonometry is the right triangle. A right triangle is a triangle where one of the angles is exactly 90°. Conduit bending layouts can often be drawn as simple right triangles. When performing calculations with right triangles, one of the angles other than the 90° angle is chosen as the reference angle. In conduit bending, the reference angle is the bend angle.

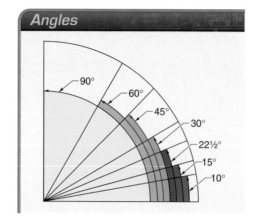

**Figure B-1.** An angle is a measure of the rotation between lines. Conduit is bent to standard angles.

The reference angle is often designated as angle theta (θ). **See Figure B-2.** Theta (θ) is a letter in the Greek alphabet often used by mathematicians to label angles. The sides of the triangle are given names based on their location relative to the reference angle.

The side of the triangle next to the reference angle and connecting to the right angle is called the adjacent side (A). The side of the angle opposite the right angle is called the hypotenuse (H). The side opposite the reference angle is called the opposite side (O).

## Trigonometry Functions

A function is a mathematical equation used to solve for unknown values. There are six common trig functions based on the right triangle. These functions are the sine (sin), cosine (cos), tangent (tan), cosecant (csc), secant (sec), and cotangent (cot) functions. **See Figure B-3.**

## Right Triangles

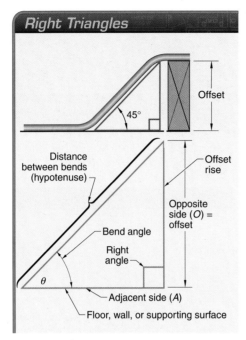

**Figure B-2.** *A right triangle is a triangle where one of the angles is exactly 90°. Right triangles are used to calculate the distance between bends.*

## Trigonometry Functions

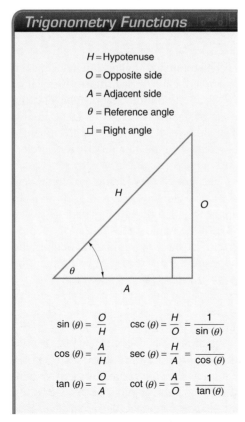

**Figure B-3.** *There are six common related trigonometry functions used with triangles.*

The functions are based on the ratios of the lengths of the sides of the triangle and are defined relative to the reference angle. For example, the sine function is the ratio of the length of the opposite side to the length of the hypotenuse. The cosecant function is the reciprocal of the sine function. These functions are conventionally written as follows:

$$\sin(\theta) = \frac{opposite}{hypotenuse} \quad \text{or} \quad \sin(\theta) = \frac{O}{H}$$

and

$$\csc(\theta) = \frac{hypotenuse}{opposite} \quad \text{or} \quad \csc(\theta) = \frac{H}{O}$$

Likewise, the cosine function is the ratio of the length of the adjacent side to the length of the hypotenuse. The secant function is the reciprocal of the cosine function. These functions are conventionally written as follows:

$$\cos(\theta) = \frac{adjacent}{hypotenuse} \quad \text{or} \quad \cos(\theta) = \frac{A}{H}$$

and

$$\sec(\theta) = \frac{hypotenuse}{adjacent} \quad \text{or} \quad \sec(\theta) = \frac{H}{A}$$

Similarly, the tangent function is the ratio of the length of the opposite side to the length of the adjacent side. The cotangent function is the reciprocal of the tangent function. These functions are conventionally written as follows:

$$\tan(\theta) = \frac{opposite}{adjacent} \quad \text{or} \quad \tan(\theta) = \frac{O}{A}$$

and

$$\cot(\theta) = \frac{adjacent}{opposite} \quad \text{or} \quad \cot(\theta) = \frac{A}{O}$$

## TRIGONOMETRY AND CONDUIT BENDING

There are some situations where the standard multipliers for the distance between bends and for shrink cannot be used, such as when a bend is made at a nonstandard angle. Trigonometry can be used to calculate the correct multipliers.

In addition, many of the calculations used with hand bending are simplifications of the actual calculations needed to pre-position bends in conduit. Values for the trigonometry functions can be found in the Appendix tables or by using a scientific calculator.

## Gain

When conduit is bent, it takes a shortcut along the wall or supporting surface. The end of the conduit seems to extend farther than the distance the conduit would follow if it were bent along the straight distance exactly as the wall is formed. **See Figure B-4.** The arc length is the distance along the curved path where the conduit is bent. Gain is the difference between the run length, including the straight distance, and the actual length of conduit, including the arc length.

**Gain for 90° Bends.** The bend radius is the radius of a circle aligned with the bend along the centerline. For a 90° bend, the lengths of the two straight sections (run) are equal to the bend radius plus half the conduit outside diameter (*OD*), or $L_S = R + \frac{1}{2} OD$. From trigonometry, the arc length is the bend radius multiplied by the angle and a conversion factor, or $L_A = \frac{R\theta\pi}{180}$. For a 90° bend, $\theta = 90°$ and $\frac{R\theta\pi}{180} = \frac{R\pi}{2}$. Gain is calculated as follows:

$$gain = 2L_S - L_A$$
$$gain = 2(R + \tfrac{1}{2} OD) - \tfrac{R\pi}{2}$$

where

$gain$ = gain from the bend, in inches
$L_S$ = length of straight distances, in inches
$L_A$ = length of arc, in inches
$R$ = bend radius, in inches
$OD$ = conduit outside diameter, in inches

This equation can be simplified by rearranging the terms of the equation. Using algebra, the equation can be rearranged as follows:

$$gain = 2(R + \tfrac{1}{2} OD) - \tfrac{R\pi}{2}$$
$$gain = 2R + OD - \tfrac{R\pi}{2}$$
$$gain = OD + R(2 - \tfrac{\pi}{2})$$
$$gain = OD + 0.4292R$$

For example, a 90° bend is to be fabricated with ¾″ EMT conduit. The bend radius for a typical ¾″ EMT hand bender is 5⅛″ (5.125″), and the conduit OD is 0.922″. The gain is calculated as follows:

$$gain = OD + 0.4292R$$
$$gain = 0.922 + (0.4292 \times 5.125)$$
$$gain = 0.922 + 2.20$$
$$gain = \mathbf{3.12''}, \text{ or } \mathbf{3\tfrac{1}{8}''}$$

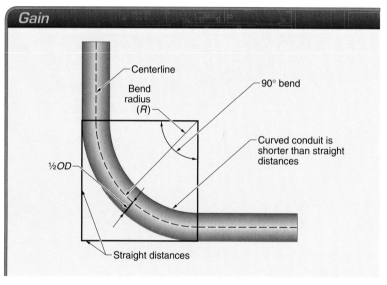

**Figure B-4.** *Gain represents the shortcut the conduit takes as it is bent into a curve.*

**Gain for Any Bend Angle.** There are occasional circumstances where the gain for a bend of any angle must be calculated. Like calculating gain for a 90° bend, the gain is the difference between the straight sections and the arc length of the bend. **See Figure B-5.**

In this case, the straight lengths are equal to the straight length of a 90° bend multiplied by the tangent of half the bend angle. The arc length is the product of the bend radius, bend angle, and a conversion factor, $\frac{\pi}{180}$. Gain for any bend angle is calculated as follows:

$$gain = 2(R + \tfrac{1}{2}OD)\tan(\tfrac{\theta}{2}) - (\tfrac{R\theta\pi}{180})$$

where

$gain$ = gain, in inches
$R$ = bend radius, in inches
$OD$ = conduit outside diameter, in inches
$\theta$ = bend angle, in degrees
$\tan(\tfrac{\theta}{2})$ = tangent of half the bend angle

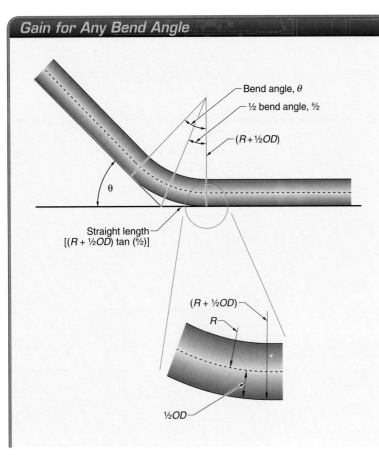

**Gain for Any Bend Angle**

Bend angle, $\theta$
½ bend angle, $\tfrac{\theta}{2}$
$(R + \tfrac{1}{2}OD)$
$\theta$
Straight length
$[(R + \tfrac{1}{2}OD)\tan(\tfrac{\theta}{2})]$

$(R + \tfrac{1}{2}OD)$
$R$
½$OD$

*Figure B-5. Gain can be calculated for any bend angle.*

For example, a 45° bend is required in a length of 2″ rigid conduit with a bend radius of 9½″ (9.50″). **See Figure B-6.** The outside diameter of 2″ rigid conduit is 2.375″. In order to cut and thread the conduit before bending, the gain must be determined so the conduit can be cut to the correct length. The gain is calculated as follows:

$$gain = 2(R + \tfrac{1}{2}OD)\tan(\tfrac{\theta}{2}) - (\tfrac{R\theta\pi}{180})$$
$$gain = 2(9.50 + \tfrac{1}{2} \times 2.375)\tan(\tfrac{45}{2}) - $$
$$(9.50 \times 45 \times \tfrac{3.14}{180})$$
$$gain = 2(9.50 + 1.19)\tan(22.5) - (7.46)$$
$$gain = 2(9.50 + 1.19) \times 0.414 - (7.46)$$
$$gain = 2(10.69) \times 0.414 - 7.46$$
$$gain = 8.86 - 7.46$$
$$gain = \mathbf{1.4″},\ or\ \mathbf{1\tfrac{3}{8}″}$$

## Distance between Bends

The distance between bends is the distance between the ends of the bends in an offset. It may also be measured from the centers of each bend. The simplified method for calculating the distance between bends does not take the gain into account. As a result, the distance between bends can end up being too large, and the offset will rise too far above the obstruction.

**Simplified Method for Distance between Bends.** The traditional method used to calculate the distance between bends is to calculate the product of the distance multiplier, $M$, and the offset rise, $O$. **See Figure B-7.** The distance multiplier is equal to $csc(\theta)$. For example, 4″ conduit is bent over a 14″ obstruction with a 45° bend angle. The simplified method for calculating the distance between bends is as follows:

$$D = csc(\theta) \times O$$
$$D = csc(45) \times 14$$
$$D = 1.414 \times 14$$
$$D = \mathbf{19.8″}$$

**Actual Distance between Bends.** The actual distance between bends is distance BD, along the dashed arc length. This is equal to the straight length ($B$ to $C$) plus the arc length ($C$ to $D$). The straight length between bends, $BC$, is calculated as follows:

$$BC = (csc(\theta) \times O) - 2R\tan\left(\frac{\theta}{2}\right)$$

where

$BC$ = straight length between bends, in inches
$\theta$ = bend angle, in degrees
$O$ = offset elevation, in inches
$R$ = bend radius, in inches

After the straight length is known, the arc length can be calculated. The arc length, $CD$, is calculated as follows:

$$CD = \frac{R\theta\pi}{180}$$

where

$CD$ = arc length, in inches
$R$ = bend radius, in inches
$\theta$ = bend angle, in degrees

Therefore, length BD is calculated as follows:

$$BD = (\csc(\theta) \times O) -$$
$$2R\tan\left(\frac{\theta}{2}\right) + \frac{R\theta\pi}{180}$$

The calculation for the actual distance between bends needs to take the gain into account. For example, 4″ conduit is bent over a 14″ obstruction with a 45° bend angle. For this type of situation, a typical bend radius is 20″. The distance between bends is calculated as follows:

$$BD = (\csc(\theta) \times O) -$$
$$2R\tan\left(\frac{\theta}{2}\right) + \frac{R\theta\pi}{180}$$

$$BD = (\csc(45) \times 14) -$$
$$(2 \times 20)\tan\left(\frac{45}{2}\right) +$$
$$\frac{20 \times 45 \times \pi}{180}$$

$BD = (1.414 \times 14) - 40\tan(22.5) + 15.708$
$BD = (1.414 \times 14) - (40 \times 0.414) + 15.708$
$BD = 19.80 - 16.56 + 15.708$
$BD = $ **18.9″**, or **18⅞″**

The distance between bends calculated by the simplified method is 19.8″. The correct distance between bends is 18.9″. Therefore, the distance between bends is nearly an inch too long. This will create a gap between the conduit and the obstruction when making bends in larger conduit.

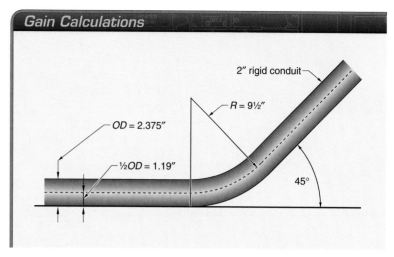

**Figure B-6.** Gain is calculated from the bend angle, bend radius, and conduit OD.

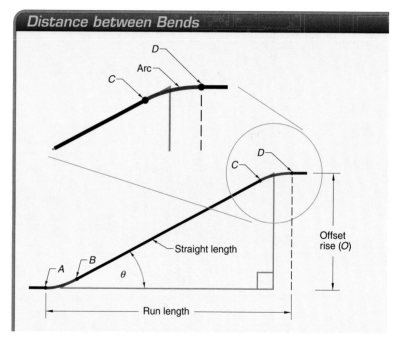

**Figure B-7.** The simplified method for calculating the distance between bends uses the offset rise and the bend angle. The complete method also uses the gain.

### Shrink

For hand bending, there are some simple calculations used to determine where to place the pencil marks on the conduit so that the bends occur in the right places. When bending large size conduit, these simple calculations are not as accurate as they need to be. Using the simplified calculations, the calculated shrink and distance between bends results in conduit that is too long after the bends are completed. This occurs because the simplified method used in hand bending does not take gain into account.

**Simplified Method for Shrink.** The simplest approximation to the shrink calculation involves using the triangle created by the bends of the offset. **See Figure B-8.** The difference between the length of the hypotenuse and the length of the adjacent side is the amount of shrink. This can be written in trigonometry form as follows:

$$S = \csc(\theta) \times O - \cot(\theta) \times O$$

or

$$S = (\csc(\theta) - \cot(\theta)) \times O$$

where

$S$ = total shrink
$\theta$ = bend angle, in degrees
$\csc(\theta)$ = cosecant of the bend angle
$\cot(\theta)$ = cotangent of the bend angle
$O$ = offset elevation

The term "$\csc(\theta) - \cot(\theta)$" is the shrink constant given in tables. For large conduit with a large bend radius, the gain needs to be taken into account. In this case, the shrink constant depends on the bend radius as well as the offset rise and the bend angle.

The simplified method calculates the shrink as the product of the shrink constant and the offset rise. The offset rise is measured in the field. The shrink constant depends on the bend angle. For small conduit used with hand benders, the simplified method produces acceptable results. However, this method results in significant inaccuracies when bending larger conduit.

For example, a 45° offset is to be bent in 4″ conduit over a 16″ obstruction. Therefore, the offset rise is 16″. From the offset shrink table, the shrink constant is 0.41. The bend angle is 45°. The shrink amount from the simplified method is calculated as follows:

$$S = c \times O$$
$$S = 0.414 \times 16$$
$$S = 6.62″, \text{ or } \mathbf{6\frac{5}{8}″}$$

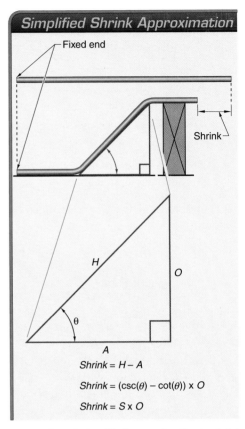

**Simplified Shrink Approximation**

Fixed end

Shrink

$H$

$O$

$\theta$

$A$

$Shrink = H - A$

$Shrink = (\csc(\theta) - \cot(\theta)) \times O$

$Shrink = S \times O$

*Figure B-8. A simplified approximation to shrink calculations uses the triangle created by the bends of the offset.*

**Actual Shrink.** The actual length is the length of the conduit that it takes to travel the run length and takes gain into account. **See Figure B-9.** The run length of conduit is the horizontal length that a conduit run covers and does not include the bends. The difference between these two lengths is the shrink.

The actual length, $L_A$, is the arc length ($A$ to $B$) + straight length ($B$ to $C$) + arc length ($C$ to $D$). The arc lengths are shown in red in the illustration. From trigonometry, we know that arc length = $\frac{R\theta\pi}{180}$ and the length $BC = O \times \csc(\theta) - 2y$. Also, $y = R\tan(\frac{\theta}{2})$. Combining these, the actual length is calculated as follows:

$$L_A = R\theta\left(\frac{\pi}{180}\right) + \csc(\theta) \times O - 2R\tan\left(\frac{\theta}{2}\right) + R\theta\left(\frac{\pi}{180}\right)$$

where

$L_A$ = actual length, in inches
$R$ = bend radius, in inches
$\theta$ = bend angle, in degrees
$O$ = offset elevation, in inches

**Tech Fact**

A bender should never be used to bend anything except EMT or rigid conduit as marked on the bender. Bending other objects, such as rebar, can permanently damage a bender so that it can no longer make accurate bends.

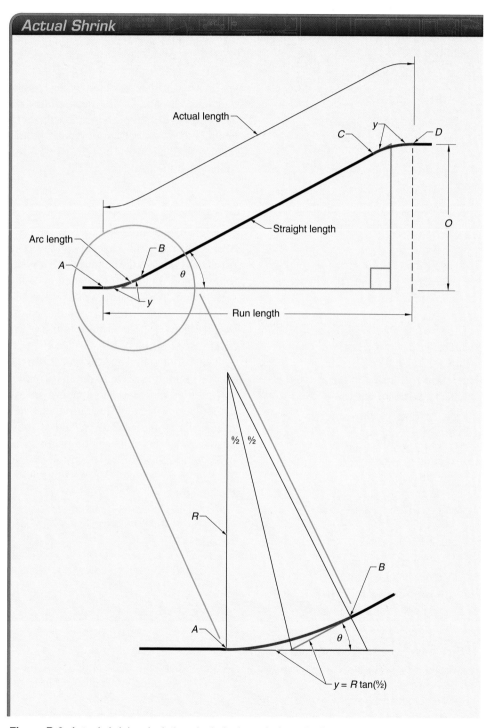

**Actual Shrink**

Actual length

Straight length

Arc length

Run length

$O$

$y = R \tan(\theta/2)$

**Figure B-9.** *Actual shrink calculations include the gain from the bends.*

The run length, $L_R$, is the distance from point $A$ along the horizontal line to a point directly below point $D$. From trigonometry, we know that the length of the horizontal section of the triangle is $\cot(\theta) \times O$. To extend from the triangle to the end of the bends, we need to add $2 \times y$. From trigonometry, we know that $y = R\tan(\theta/2)$. Combining these, the run length is calculated as follows:

$$L_R = \cot(\theta) \times O + 2R\tan\left(\frac{\theta}{2}\right)$$

where

$L_R$ = run length, in inches
$\theta$ = bend angle, in degrees
$O$ = offset elevation, in inches
$R$ = bend radius, in inches

Since shrink is the difference between the actual length and the run length, it is calculated as follows:

$$S = L_A - L_R$$

$$S = R\theta\left(\frac{\pi}{180}\right) + \csc(\theta) \times O -$$

$$2R\tan\left(\frac{\theta}{2}\right) + R\theta\left(\frac{\pi}{180}\right) -$$

$$\left(\cot(\theta) \times O + 2R\tan\left(\frac{\theta}{2}\right)\right)$$

This simplifies as follows:

$$S = (\csc(\theta) - \cot(\theta)) \times$$

$$O + \frac{2R\theta\pi}{180} - 4R(c)$$

or

$$S = (c)\,O + \frac{2R\theta\pi}{180} - 4R(c)$$

where

$S$ = shrink, in inches
$c$ = shrink constant
$O$ = offset rise, in inches
$R$ = bend radius, in inches
$\theta$ = bend angle, in degrees

In order to calculate the actual amount of shrink, the bend radius must also be known. For this situation, a typical bend radius is 20″. This amount can vary quite a bit for different benders. The shrink amount from the complete method is calculated as follows:

$$S = (c)O + \frac{2R\theta\pi}{180} - 4R(c)$$

$$S = 0.414 \times 16 +$$

$$\frac{2 \times 20 \times 45 \times 3.1416}{180} -$$

$$4 \times 20 \times 0.414$$

$$S = 6.62 + 31.416 - 33.12$$

$$S = \mathbf{4.92″}, \text{ or } \mathbf{4^{15}\!/_{16}″}$$

The actual shrink for this offset bend is 4.92″. The shrink calculated by the simplified method is 6.62″. This means that when the simplified method is used, a piece of conduit that was cut and threaded before bending will end up a little over $1^{11}\!/_{16}″$ too long when it is installed.

## Parallel Offsets

Parallel offsets are a set of two or more offsets fabricated in adjacent conduit. The parallel sections of the conduits may appear to be crowded in the angles section of the offset bend. When parallel offsets are visible, a little extra work is required to adjust the layout to improve the appearance of the bends. **See Figure B-10.**

The layout marks must be moved on the second conduit relative to the first conduit in order to increase the spacing in the angled section. In the illustration, distance $D$ is the distance between centerlines. Distance $A$ represents the adjustment, or distance to move the bends in the second conduit, to equalize the distances between the conduits.

Using trigonometry on the small triangle at the top, the distance "$x$" can be calculated as follows:

$$\sin(\theta) = \frac{D - D\cos(\theta)}{A}$$

or $\quad A = \dfrac{D(1 - \cos(\theta))}{\sin(\theta)}$

where

$\theta$ = bend angle, in degrees
$D$ = distance between conduit centerlines, in inches
$A$ = adjustment to increase spacing, in inches

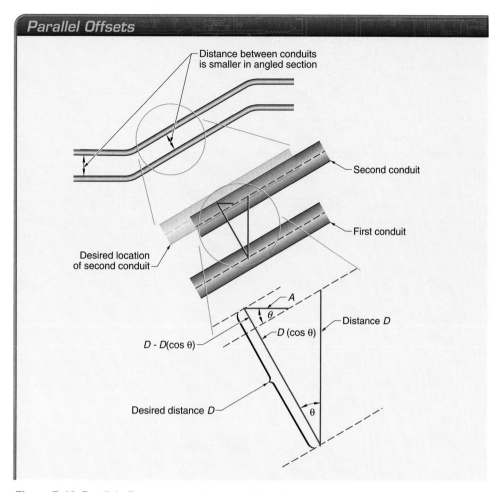

**Parallel Offsets**

*Figure B-10. Parallel offsets appear to be crowded unless the layout is adjusted.*

The half-angle identity for tangent is

$$\tan\left(\frac{\theta}{2}\right) = \frac{1 - \cos(\theta)}{\sin(\theta)}$$

Substitute and rearrange to

$$A = D \tan\left(\frac{\theta}{2}\right)$$

The amount that the pencil marks must be shifted can be calculated from this formula. For example, a rack of conduit is coming out of a junction box. The conduits are spaced 3″ apart on center with a 30° offset. The adjustment required in placing the pencil marks is calculated as follows:

$$A = D \tan\left(\frac{\theta}{2}\right)$$

$$A = 3 \tan\left(\frac{30}{2}\right)$$

$$A = 3 \tan(15)$$
$$A = 3 \times 0.268$$
$$A = \mathbf{0.804''}$$

It should be noted that the multiplier for the adjustment for parallel offsets is the same multiplier as for shrink. This can be shown as follows:

$$\tan\left(\frac{\theta}{2}\right) = \frac{1 - \cos(\theta)}{\sin(\theta)}$$

$$\frac{1 - \cos(\theta)}{\sin(\theta)} = \frac{1}{\sin(\theta)} - \frac{\cos(\theta)}{\sin(\theta)}$$

$$\frac{1}{\sin(\theta)} - \frac{\cos(\theta)}{\sin(\theta)} = \csc(\theta) - \cot(\theta)$$

The term "$\csc(\theta) - \cot(\theta)$" is the shrink constant used earlier.

The equation to calculate shrink was determined to be as follows:

$$S = R\theta\left(\frac{\pi}{180}\right) + \csc(\theta) \times O - 2R\tan\left(\frac{\theta}{2}\right) +$$

$$R\theta\left(\frac{\pi}{180}\right) - (\cot(\theta) \times O + 2R\tan\left(\frac{\theta}{2}\right))$$

This equation is complex and difficult to use. It can be simplified by rearranging the terms to give the following:

$$S = \frac{2R\theta\pi}{180} + (\csc(\theta) - \cot(\theta)) \times O -$$

$$4R\tan\left(\frac{\theta}{2}\right)$$

Use the half angle identity for tangent, tan(θ/2) = (1-cos(θ))/sin(θ), as follows:

$$S = \frac{2R\theta\pi}{180} + (\csc(\theta) - \cot(\theta)) \times O -$$

$$4R\left(\frac{1 - \cos(\theta)}{\sin(\theta)}\right)$$

Use the reciprocal identities on the quantities in the parentheses on the right as follows:

$$S = \frac{2R\theta\pi}{180} + (\csc(\theta) - \cot(\theta)) \times O -$$

$$4R\left(\frac{1}{\sin(\theta)} - \frac{\cos(\theta)}{\sin(\theta)}\right)$$

$$S = (\csc(\theta) - \cot(\theta)) \times O + \frac{2R\theta\pi}{180} -$$

$$4R(\csc(\theta) - \tan(\theta))$$

The quantity within the parentheses, $\csc(\theta) - \cot(\theta)$, is the same shrink constant determined from the simple triangle. Let $c$ represent this constant as follows:

$$S = (c)O + \frac{2R\theta\pi}{180} - 4R(c)$$

The first term in this equation is the shrink that is calculated for hand bends. The second term is the length of the two bend arcs. The third term is a gain adjustment. The bend radius is usually given in the bender manual.

## FINDING AN UNKNOWN RADIUS

There are situations where the radius of a circular or cylindrical object is not known and must be determined. This comes up when a segmented bend is to be made around an object and the radius is not known.

The simplest method to determine an unknown radius is to look at the manufacturer's literature for the object and obtain the size of the object from a drawing. A drawing will typically show the diameter of the object. The radius is exactly half the diameter.

If there is no drawing available, the unknown radius must be measured. There are several methods used to make this measurement. The straightedge method is quick and easy and requires minimal calculations. The circumference method is also fairly quick and requires only simple division to determine the radius. The trigonometry method is somewhat longer and requires more calculations.

### Straightedge Method

An easy method of finding an unknown radius is to measure it directly with a rule, tape measure, or straightedge. This method is effective when the object is cylindrical, such as a storage tank, without any obstructions across the top. The diameter of the object passes through the center of the top and is the maximum distance across the top. The straightedge is moved back and forth across the center of the top to find the maximum distance.

If a rule or tape measure is used, the distance can be measured directly. If a straightedge is used, a mark is placed on the straightedge at the point where it crosses the other edge of the cylindrical object. The distance from the end of the straightedge to the mark is the diameter.

Some tanks or vessels have rounded corners. In this case, a short extension can be used to extend the edge up to a point where the distance can be measured. If there is a small object blocking the center of the top, such as a motor, a short extension can be used to extend the edge up to a point where the straightedge can pass directly over the center. **See Figure B-11.**

For example, conduit is being run around a cylindrical chemical mixing and storage tank to connect two control box enclosures mounted on the tank. The maximum distance across the top of the tank is measured to be 12′-2 ½″. The radius of the tank is half of that, or 6′-1¼″.

## Circumference Method

There are situations where the straightedge method cannot be used to measure an unknown radius. These situations may occur when there is a large object blocking the center of the top of the tank or when the top is not accessible. If the straightedge cannot pass directly over the center, the diameter cannot be directly measured. This large object can be a motor for a mixer, an access point into the tank, or another large obstruction.

When a straightedge cannot be used to measure the diameter directly, the circumference can often be measured and the radius calculated from the circumference. The circumference can be measured by wrapping string or flexible conductor wire around the outside of the tank.

The length of string or wire that reaches entirely around the tank represents the circumference. The string can be stretched out on the ground and the length measured with a tape measure or rule. Any other flexible object can be used if string is not available.

Once the circumference is known, the radius can be calculated. As follows, the circumference of a circle is the product of the radius and $2\pi$:

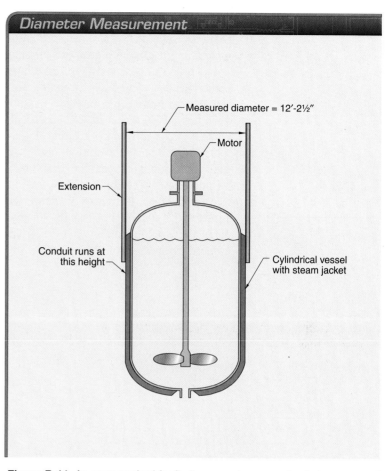

*Diameter Measurement*

Measured diameter = 12′-2½″

Motor

Extension

Conduit runs at this height

Cylindrical vessel with steam jacket

**Figure B-11.** *An easy method for finding an unknown radius is to measure it directly with a rule, tape measure, or straightedge.*

$C = 2\pi R$

where

$C$ = circumference, in inches
$R$ = radius, in inches

Therefore, the radius can be calculated by dividing the circumference by $2\pi$. For example, a circumference is measured to be 24′-6″ (24.5′). The unknown radius is calculated as follows:

$$R = \frac{C}{2\pi}$$

$$R = \frac{24.5}{2 \times 3.14}$$

$$R = \frac{24.5}{6.28}$$

$R =$ **3.90′**, or **46.8″**

## Trigonometry Method

There are situations where neither the diameter nor the circumference of a circular tank can be directly measured. This may happen when a tank passes through a wall into another room. This may also happen when the curved object is not cylindrical, but consists of only part of a circle. Trigonometry can be used to find an unknown radius by using an arc and a circle to represent the top of the tank.

An arc is any section of a circle. **See Figure B-12.** A chord is a straight line that connects two points on an arc or a circle. A chord can pass through any part of the circle, not just through the center.

The height of the arc (HA) is the farthest distance from a chord to its arc. The height of an arc is always at a 90° angle from the center of the chord.

The chord of half the arc (CHA) is a second chord, connecting one end of the original chord with the height of the arc. The chord of half the arc is the hypotenuse of a triangle formed by half the original chord and the height of the arc.

The radius of a circle can be calculated from the chord of half the arc and the height of the arc as follows:

$$R = \frac{CHA^2}{2HA}$$

where

$R$ = radius, in inches
$CHA$ = chord of half the arc, in inches
$HA$ = height of the arc, in inches

This equation is difficult to use because the chord of half the arc is hard to measure accurately. From the Pythagorean theorem, the equation can be modified as follows:

$$CHA^2 = HA^2 + (\tfrac{1}{2})^2$$

and

$$R = \frac{HA^2 + (\tfrac{1}{2})^2}{2HA}$$

where

$R$ = radius, in inches
$CHA$ = chord of half the arc, in inches
$HA$ = height of the arc, in inches
$L$ = length of the chord, in inches

The measurements for the radius, chord, and height may be in feet, inches, or mixed feet and inches. However, all the variables must be calculated in the same units, the easiest being inches. This method can be used to measure both inside and outside arcs.

**Radius of Inside Arcs.** This formula is much easier to use than it may seem. From this equation, the only required measurements are the known length of the straight-edge and the height of the arc (the distance

**Finding an Unknown Radius from a Chord**

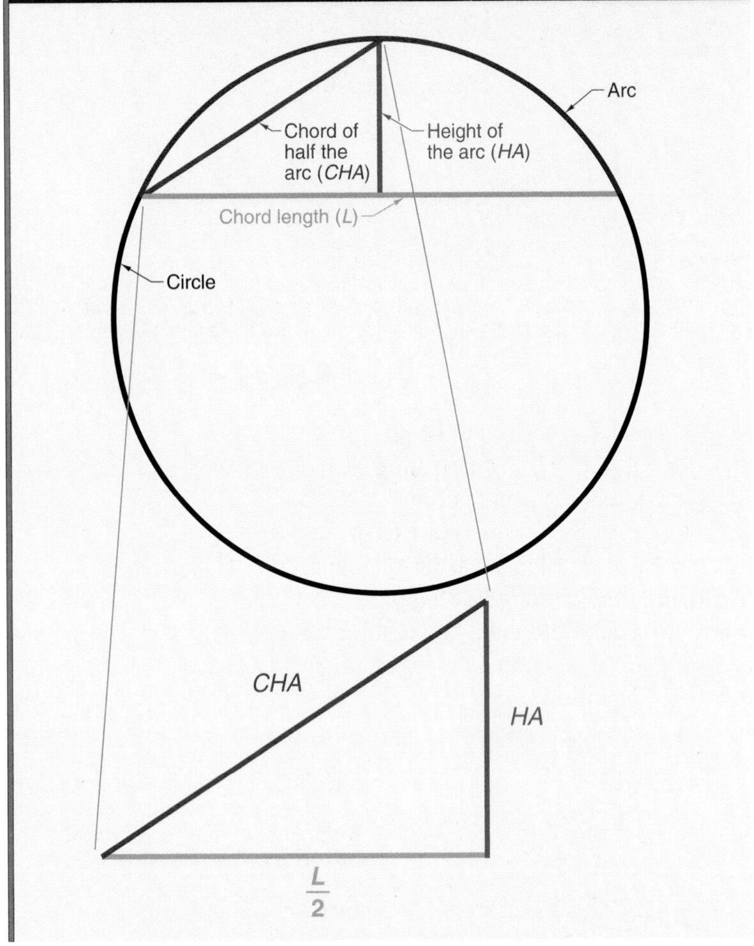

*Figure B-12. Trigonometry can be used to find the radius of a circle by measuring a chord and the height of the arc.*

from the center of the straightedge to the arc). The procedure to use this method is as follows:

1. Select a straightedge of known length, such as a measured piece of conduit. The straightedge must be smaller than the diameter. Measure and mark the center of the straightedge.

2. Place the straightedge across the top of the object so that both ends are at the edge of the arc. Measure the height of the arc. This is the length of a line at 90° from the center of the straightedge to the edge of the circle.

For example, a 10′ (120″) piece of conduit is held to the inside of a structure having an unknown arc. **See Figure B-13.** The measured height of this arc is 34″. The radius is calculated as follows:

$$R = \frac{HA^2 + (\frac{L}{2})^2}{2HA}$$

$$R = \frac{34^2 + (\frac{120}{2})^2}{2 \times 34}$$

$$R = \frac{34^2 + 60^2}{68}$$

$$R = \frac{1156 + 3600}{68}$$

$$R = \frac{4756}{68}$$

$$R = \textbf{69.94″, or 69}^{15}\!/_{16}″$$

**Radius of Outside Arcs.** The trigonometry method also works equally well on outside arcs, although the mechanics change slightly. Again, the only required measurements are the known length of the straightedge and the height of the arc. **See Figure B-14.** The procedure to use this method is as follows:

1. Choose and mark a straightedge as was done for measuring an inside arc.

2. Place the straightedge against the outside of the arc with the center mark against the edge. Measure the distance from the endpoints to the arc. Keeping the center mark on the arc, move the straightedge until the measurements from each end of the straightedge to the arc are equal.

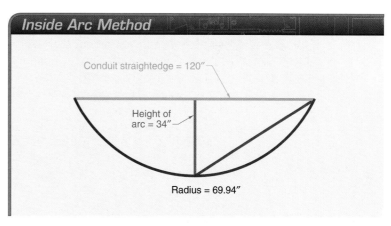

**Figure B-13.** *For the inside arc method of finding a radius, a straightedge is used as a chord and the height of the arc is measured.*

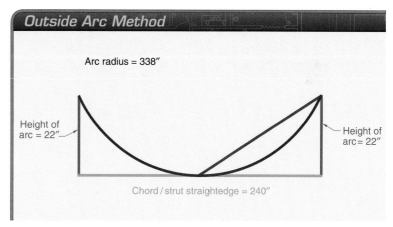

**Figure B-14.** *For the outside arc method of finding a radius, a straightedge is placed outside the circle and the height is measured at the ends of the straightedge.*

The measurement is the height of the arc. The calculation is then performed in the usual manner. For example, an outside arc is measured using a 20′ (240″) piece of strut as a straightedge. The height of the arc is 22″. The radius is calculated as follows:

$$R = \frac{HA^2 + (\frac{L}{2})^2}{2HA}$$

$$R = \frac{22^2 + (\frac{240}{2})^2}{2 \times 22}$$

$$R = \frac{22^2 + 120^2}{44}$$

$$R = \frac{484 + 14,400}{44}$$

$$R = \textbf{338″}$$

# Appendix C:
# Conduit Bending Supplements

| Trigonometry Table | | | | | | | |
|---|---|---|---|---|---|---|---|
| Angle | Sine | Cosine | Tangent | Angle | Sine | Cosine | Tangent |
| 0° | .0000 | 1.0000 | ∞ | 45° | .7071 | .7071 | 1.000 |
| 1° | .0175 | .9998 | .0175 | 46° | .7193 | .6947 | 1.0355 |
| 2° | .0349 | .9994 | .0349 | 47° | .7314 | .6820 | 1.0724 |
| 3° | .0523 | .9986 | .0524 | 48° | .7431 | .6691 | 1.1106 |
| 4° | .0698 | .9976 | .0699 | 49° | .7547 | .6561 | 1.1504 |
| 5° | .0872 | .9962 | .0875 | 50° | .7660 | .6428 | 1.1918 |
| 6° | .1045 | .9945 | .1051 | 51° | .7771 | .6293 | 1.2349 |
| 7° | .1219 | .9925 | .1228 | 52° | .7880 | .6157 | 1.2799 |
| 8° | .1392 | .9903 | .1405 | 53° | .7986 | .6018 | 1.3270 |
| 9° | .1564 | .9877 | .1584 | 54° | .8090 | .5878 | 1.3764 |
| 10° | .1736 | .9848 | .1763 | 55° | .8192 | .5736 | 1.4281 |
| 11° | .1908 | .9816 | .1944 | 56° | .8290 | .5592 | 1.4826 |
| 12° | .2079 | .9781 | .2126 | 57° | .8387 | .5446 | 1.5399 |
| 13° | .2250 | .9744 | .2309 | 58° | .8480 | .5299 | 1.6003 |
| 14° | .2419 | .9703 | .2493 | 59° | .8572 | .5150 | 1.6643 |
| 15° | .2588 | .9659 | .2679 | 60° | .8660 | .5000 | 1.7321 |
| 16° | .2756 | .9613 | .2867 | 61° | .8746 | .4848 | 1.8040 |
| 17° | .2924 | .9563 | .3057 | 62° | .8829 | .4695 | 1.8807 |
| 18° | .3090 | .9511 | .3249 | 63° | .8910 | .4540 | 1.9626 |
| 19° | .3256 | .9455 | .3443 | 64° | .8988 | .4384 | 2.0503 |
| 20° | .3420 | .9397 | .3640 | 65° | .9063 | .4226 | 2.1445 |
| 21° | .3584 | .9336 | .3839 | 66° | .9135 | .4067 | 2.2460 |
| 22° | .3746 | .9272 | .4040 | 67° | .9205 | .3907 | 2.3559 |
| 22½° | .3827 | .9239 | .4142 | 68° | .9272 | .3746 | 2.4751 |
| 23° | .3907 | .9205 | .4245 | 69° | .9336 | .3584 | 2.6051 |
| 24° | .4067 | .9135 | .4452 | 70° | .9397 | .3420 | 2.7475 |
| 25° | .4226 | .9063 | .4663 | 71° | .9455 | .3256 | 2.9042 |
| 26° | .4384 | .8988 | .4877 | 72° | .9511 | .3090 | 3.0777 |
| 27° | .4540 | .8910 | .5095 | 73° | .9563 | .2924 | 3.2709 |
| 28° | .4695 | .8829 | .5317 | 74° | .9613 | .2756 | 3.4874 |
| 29° | .4848 | .8746 | .5543 | 75° | .9659 | .2588 | 3.7321 |
| 30° | .5000 | .8660 | .5774 | 76° | .9703 | .2419 | 4.0108 |
| 31° | .5150 | .8572 | .6009 | 77° | .9744 | .2250 | 4.3315 |
| 32° | .5229 | .8480 | .6249 | 78° | .9781 | .2079 | 4.7046 |
| 33° | .5446 | .8387 | .6494 | 79° | .9816 | .1908 | 5.1446 |
| 34° | .5592 | .8290 | .6745 | 80° | .9848 | .1736 | 5.6713 |
| 35° | .5736 | .8192 | .7002 | 81° | .9877 | .1564 | 6.3138 |
| 36° | .5878 | .8090 | .7265 | 82° | .9903 | .1392 | 7.1154 |
| 37° | .6018 | .7986 | .7536 | 83° | .9925 | .1219 | 8.1443 |
| 38° | .6157 | .7880 | .7813 | 84° | .9945 | .1045 | 9.5144 |
| 39° | .6293 | .7771 | .8098 | 85° | .9962 | .0872 | 11.4301 |
| 40° | .6428 | .7660 | .8391 | 86° | .9976 | .0698 | 14.3007 |
| 41° | .6561 | .7547 | .8693 | 87° | .9986 | .0523 | 19.0811 |
| 42° | .6691 | .7431 | .9004 | 88° | .9994 | .0349 | 28.6363 |
| 43° | .6820 | .7314 | .9325 | 89° | .9998 | .0175 | 57.2900 |
| 44° | .6947 | .7193 | .9657 | 90° | 1.000 | .0000 | ∞ |

## Offset Multiplier (Cosecant) Table

| Angle* | Multiplier (cosecant) | Angle* | Multiplier (cosecant) |
|---|---|---|---|
| 0 | ∞ | 45 | 1.4142 |
| 1 | 57.2987 | 46 | 1.3902 |
| 2 | 28.6537 | 47 | 1.3673 |
| 3 | 19.1073 | 48 | 1.3456 |
| 4 | 14.3356 | 49 | 1.3250 |
| 5 | 11.4737 | 50 | 1.3054 |
| 6 | 9.5668 | 51 | 1.2868 |
| 7 | 8.2055 | 52 | 1.2690 |
| 8 | 7.1853 | 53 | 1.2521 |
| 9 | 6.3925 | 54 | 1.2361 |
| 10 | 5.7588 | 55 | 1.2208 |
| 11 | 5.2408 | 56 | 1.2062 |
| 12 | 4.8097 | 57 | 1.1924 |
| 13 | 4.4454 | 58 | 1.1792 |
| 14 | 4.1336 | 59 | 1.1666 |
| 15 | 3.8637 | 60 | 1.1547 |
| 16 | 3.6280 | 61 | 1.1434 |
| 17 | 3.4203 | 62 | 1.1326 |
| 18 | 3.2361 | 63 | 1.1223 |
| 19 | 3.0716 | 64 | 1.1126 |
| 20 | 2.9238 | 65 | 1.1034 |
| 21 | 2.7904 | 66 | 1.0946 |
| 22 | 2.6695 | 67 | 1.0864 |
| 22½ | 2.6131 | 68 | 1.0785 |
| 23 | 2.5593 | 69 | 1.0711 |
| 24 | 2.4586 | 70 | 1.0642 |
| 25 | 2.3662 | 71 | 1.0576 |
| 26 | 2.2812 | 72 | 1.0515 |
| 27 | 2.2027 | 73 | 1.0457 |
| 28 | 2.1301 | 74 | 1.0403 |
| 29 | 2.0627 | 75 | 1.0353 |
| 30 | 2.0000 | 76 | 1.0306 |
| 31 | 1.9416 | 77 | 1.0263 |
| 32 | 1.8871 | 78 | 1.0223 |
| 33 | 1.8361 | 79 | 1.0187 |
| 34 | 1.7883 | 80 | 1.0154 |
| 35 | 1.7434 | 81 | 1.0125 |
| 36 | 1.7013 | 82 | 1.0098 |
| 37 | 1.6616 | 83 | 1.0075 |
| 38 | 1.6243 | 84 | 1.0055 |
| 39 | 1.5890 | 85 | 1.0038 |
| 40 | 1.5557 | 86 | 1.0024 |
| 41 | 1.5243 | 87 | 1.0014 |
| 42 | 1.4945 | 88 | 1.0006 |
| 43 | 1.4663 | 89 | 1.0002 |
| 44 | 1.4396 | 90 | 1.0000 |

* in degrees

## Bending Quick Guide

### Take-up

| Bender Size | Typical Take-up |
|---|---|
| ½″ | 5″ |
| ¾″ | 6″ |
| 1″ | 8″ |
| 1¼″ | 11″ |

### Hand Bending Gain

| Conduit Size | Typical EMT Bend Radius | EMT Gain | Typical Rigid Bend Radius | Rigid Gain |
|---|---|---|---|---|
| ½″ | 4³⁄₁₆″ | 2½″ | 5⅛″ | 3¹⁄₁₆″ |
| ¾″ | 5⅛″ | 3″ | 6½″ | 3¹³⁄₁₆″ |
| 1″ | 6½″ | 3¹⁵⁄₁₆″ | 9⅝″ | 5⁷⁄₁₆″ |
| 1¼″ | 8″ | 4¹⁵⁄₁₆″ | – | – |
|  | 9⅝″ | 5⅝″ | – | – |

### Distance Multiplier and Shrink Constant

| Bend Angle, θ | Distance Multiplier | Shrink Constant |
|---|---|---|
| 5° | 11.4 | 0.044 |
| 10° | 5.76 | 0.087 |
| 15° | 3.86 | 0.13 |
| 22½° | 2.61 | 0.20 |
| 30° | 2.00 | 0.27 |
| 45° | 1.41 | 0.41 |

### Three-Bend Saddle Multiplier

| Side Bend Angle | Distance Multiplier |
|---|---|
| 15° | 3.8 |
| 22½° | 2.5 |

## Conduit Dimensions

### GRC Rigid

| Trade Size | ½ | ¾ | 1 | 1¼ | 1½ | 2 | 2½ | 3 | 3½ | 4 | 5 | 6 |
|---|---|---|---|---|---|---|---|---|---|---|---|---|
| O D | .840 | 1.050 | 1.315 | 1.660 | 1.900 | 2.375 | 2.875 | 3.500 | 4.000 | 4.500 | 5.563 | 6.625 |
| I D | .622 | .824 | 1.049 | 1.380 | 1.610 | 2.067 | 2.469 | 3.068 | 3.548 | 4.026 | 5.047 | 6.065 |
| Wall | .109 | .113 | .133 | .140 | .145 | .154 | .203 | .216 | .226 | .237 | .258 | .280 |
| Weight per 10′ | 8.2 | 11 | 16 | 21 | 26 | 35 | 56 | 73 | 88 | 103 | 140 | 184 |

### IMC

| Trade Size | ½ | ¾ | 1 | 1¼ | 1½ | 2 | 2½ | 3 | 3½ | 4 |
|---|---|---|---|---|---|---|---|---|---|---|
| O D | .815 | 1.029 | 1.290 | 1.638 | 1.883 | 2.360 | 2.857 | 3.476 | 3.971 | 4.466 |
| I D | .675 | .879 | 1.120 | 1.468 | 1.703 | 2.170 | 2.597 | 3.216 | 3.711 | 4.206 |
| Wall | .070 | .075 | .085 | .090 | .090 | .095 | .130 | .130 | .130 | .130 |
| Weight per 10′ | 6.2 | 8.4 | 12 | 16 | 19 | 26 | 44 | 54 | 63 | 70 |

### EMT

| Trade Size | ½ | ¾ | 1 | 1¼ | 1½ | 2 | 2½ | 3 | 4 |
|---|---|---|---|---|---|---|---|---|---|
| O D | .706 | .922 | 1.163 | 1.510 | 1.740 | 2.197 | 2.875 | 3.500 | 4.500 |
| I D | .622 | .824 | 1.049 | 1.380 | 1.610 | 2.067 | 2.731 | 3.356 | 4.334 |
| Wall | .042 | .049 | .057 | .065 | .065 | .065 | .072 | .072 | .083 |
| Weight per 10′ | 3.0 | 4.6 | 6.7 | 10 | 12 | 15 | 22 | 26 | 39 |

### PVC Schedule 20

| Trade Size | ½ | ¾ | 1 | 1¼ | 1½ | 2 | 2½ | 3 | 4 |
|---|---|---|---|---|---|---|---|---|---|
| O D | .840 | 1.050 | 1.315 | 1.660 | 1.900 | 2.375 | 2.875 | 3.500 | 4.500 |
| I D | .720 | .930 | 1.195 | 1.520 | 1.740 | 2.175 | 2.655 | 3.250 | 4.200 |
| Wall | .060 | .060 | .060 | .070 | .080 | .100 | .110 | .0125 | .150 |
| Weight per 10′ | 1.1 | 1.3 | 1.7 | 2.5 | 3.2 | 5 | 6.7 | 9 | 14.6 |

### PVC Schedule 40

| Trade Size | ½ | ¾ | 1 | 1¼ | 1½ | 2 | 2½ | 3 | 3½ | 4 | 5 | 6 |
|---|---|---|---|---|---|---|---|---|---|---|---|---|
| O D | .840 | 1.050 | 1.315 | 1.660 | 1.900 | 2.375 | 2.875 | 3.500 | 4.000 | 4.500 | 5.563 | 6.625 |
| I D | .622 | .824 | 1.049 | 1.380 | 1.610 | 2.067 | 2.469 | 3.068 | 3.548 | 4.026 | 5.047 | 6.065 |
| Wall | .109 | .113 | .133 | .140 | .145 | .154 | .203 | .216 | .226 | .237 | .258 | .280 |
| Weight per 10′ | 1.6 | 2.2 | 3.2 | 4.3 | 5.2 | 6.9 | 11 | 14 | 17 | 20 | 27 | 35 |

### PVC Schedule 80

| Trade Size | ½ | ¾ | 1 | 1¼ | 1½ | 2 | 2½ | 3 | 3½ | 4 | 5 | 6 |
|---|---|---|---|---|---|---|---|---|---|---|---|---|
| O D | .840 | 1.050 | 1.315 | 1.660 | 1.900 | 2.375 | 2.875 | 3.500 | 4.000 | 4.500 | 5.563 | 6.625 |
| I D | .546 | .742 | .957 | 1.278 | 1.500 | 1.939 | 2.323 | 2.900 | 3.364 | 3.826 | 4.813 | 5.761 |
| Wall | .147 | .154 | .179 | .191 | .200 | .218 | .276 | .300 | .318 | .337 | .375 | .432 |
| Weight per 10′ | 2.1 | 2.8 | 4.1 | 5.7 | 6.8 | 9.4 | 14 | 19 | 23 | 28 | 39 | 54 |

### Aluminum Rigid

| Trade Size | ½ | ¾ | 1 | 1¼ | 1½ | 2 | 2½ | 3 | 3½ | 4 | 5 | 6 |
|---|---|---|---|---|---|---|---|---|---|---|---|---|
| O D | 0.840 | 1.050 | 1.315 | 1.660 | 1.900 | 2.375 | 2.875 | 3.500 | 4.000 | 4.500 | 5.563 | 6.625 |
| I D | 0.632 | 0.836 | 1.063 | 1.394 | 1.624 | 2.083 | 2.489 | 3.090 | 3.570 | 4.050 | 5.073 | 6.093 |
| Wall | 0.104 | 0.107 | 0.126 | 0.133 | 0.138 | 0.146 | 0.193 | 0.205 | 0.215 | 0.225 | 0.245 | 0.266 |
| Weight per 10′ | 2.8 | 3.7 | 5.5 | 7.2 | 8.9 | 12 | 19 | 25 | 30 | 35 | 48 | 63 |

### Aluminum EMT

| Trade Size | 2 | 2½ | 3 | 3½ | 4 |
|---|---|---|---|---|---|
| O D | 2.197 | 2.875 | 3.500 | 4.000 | 4.500 |
| I D | 2.051 | 2.711 | 3.332 | 3.810 | 4.304 |
| Wall | 0.073 | 0.082 | 0.084 | 0.095 | 0.098 |
| Weight per 10′ | 5.7 | 8.5 | 11 | 14 | 16 |

## Conduit Layout Guide

### Minimum Center-to-Center Measurement*

| | ½ | ¾ | 1 | 1¼ | 1½ | 2 | 2½ | 3 | 3½ | 4 | 5 | 6 |
|---|---|---|---|---|---|---|---|---|---|---|---|---|
| ½ | .840 | .954 | 1.078 | 1.250 | 1.370 | 1.608 | 1.0858 | 2.170 | 2.420 | 2.670 | 3.202 | 3.733 |
| ¾ | .954 | 1.050 | 1.183 | 1.355 | 1.475 | 1.713 | 1.963 | 2.275 | 2.525 | 2.775 | 3.307 | 3.838 |
| 1 | 1.078 | 1.183 | 1.315 | 1.488 | 1.608 | 1.846 | 2.096 | 2.408 | 2.658 | 2.908 | 3.440 | 3.971 |
| 1¼ | 1.250 | 1.355 | 1.488 | 1.660 | 1.780 | 2.018 | 2.268 | 2.580 | 2.830 | 3.080 | 3.612 | 4.143 |
| 1½ | 1.370 | 1.475 | 1.608 | 1.780 | 1.900 | 2.138 | 2.388 | 2.700 | 2.950 | 3.200 | 3.732 | 4.263 |
| 2 | 1.608 | 1.713 | 1.846 | 2.018 | 2.138 | 2.375 | 2.626 | 2.938 | 3.188 | 3.438 | 3.970 | 4.501 |
| 2½ | 1.858 | 1.963 | 2.096 | 2.268 | 2.388 | 2.626 | 2.876 | 3.188 | 3.438 | 3.688 | 4.220 | 4.751 |
| 3 | 2.170 | 2.275 | 2.408 | 2.580 | 2.700 | 2.938 | 3.188 | 3.500 | 3.750 | 4.000 | 4.532 | 5.063 |
| 3½ | 2.420 | 2.525 | 2.658 | 2.830 | 2.950 | 3.188 | 3.438 | 3.750 | 4.000 | 4.250 | 4.782 | 5.313 |
| 4 | 2.670 | 2.775 | 2.908 | 3.080 | 3.200 | 3.438 | 3.688 | 4.00 | 4.250 | 4.500 | 5.032 | 5.563 |
| 5 | 3.202 | 3.307 | 3.440 | 3.612 | 3.732 | 3.970 | 4.220 | 4.532 | 4.782 | 5.032 | 5.564 | 6.095 |
| 6 | 3.733 | 3.838 | 3.971 | 4.143 | 4.263 | 4.501 | 4.751 | 5.063 | 5.313 | 5.563 | 6.095 | 6.625 |

* in inches

### Conduit Bushing Diameters*

| Bushing Trade Size | Bushing Diameter |
|---|---|
| ½ | 1 1/32 |
| ¾ | 1¼ |
| 1 | 1 17/32 |
| 1¼ | 1 20/32 |
| 1½ | 2 5/32 |
| 2 | 2 11/16 |
| 2½ | 3 5/32 |
| 3 | 3 25/32 |

* in inches

### Conduit Locknut Diameters*

| Locknut Trade Size | Locknut Diameter |
|---|---|
| ½ | 1 1/16 |
| ¾ | 1 7/16 |
| 1 | 1 23/32 |
| 1¼ | 1 5/16 |
| 1½ | 2 9/16 |
| 2 | 3 1/8 |
| 2½ | 3¾ |
| 3 | 4 |

* in inches

### Expansion Characteristics of Steel Conduit and Tubing

Coefficient of Thermal Expansion = $6.5 \times 10^{-6}$ in./in./ °F

| Temperature Changes in °F | Length Change in in./100 ft of Steel Conduit | Temperature Changes in °F | Length Change in in./100 ft of Steel Conduit | Temperature Changes in °F | Length Change in in./100 ft of Steel Conduit | Temperature Changes in °F | Length Change in in./100 ft of Steel Conduit |
|---|---|---|---|---|---|---|---|
| 5 | 0.04 | 55 | 0.44 | 105 | 0.84 | 155 | 126 |
| 10 | 0.08 | 60 | 0.48 | 110 | 0.90 | 160 | 1.30 |
| 15 | 0.12 | 65 | 0.52 | 115 | 0.94 | 165 | 1.34 |
| 20 | 0.16 | 70 | 0.56 | 120 | 0.98 | 170 | 1.38 |
| 25 | 0.20 | 75 | 0.60 | 125 | 1.02 | 175 | 1.42 |
| 30 | 0.24 | 80 | 0.64 | 130 | 1.06 | 180 | 1.46 |
| 35 | 0.28 | 85 | 0.68 | 135 | 1.10 | 185 | 1.50 |
| 40 | 0.32 | 90 | 0.72 | 140 | 1.14 | 190 | 1.54 |
| 45 | 0.36 | 05 | 0.76 | 145 | 1.18 | 195 | 1.58 |
| 50 | 0.40 | 100 | 0.82 | 150 | 1.22 | 200 | 1.62 |

*Steel Tube Institute of North America*

## Decimal Feet and Inches Conversions

| Decimal Feet | Inches |
|:---:|:---:|
| 0.05 | 5/8 |
| 0.1 | 1 3/16 |
| 0.15 | 1 13/16 |
| 0.2 | 2 3/8 |
| 0.3 | 3 5/8 |
| 0.4 | 4 13/16 |
| 0.5 | 6 |
| 0.6 | 7 3/16 |
| 0.7 | 8 3/8 |
| 0.8 | 9 5/8 |
| 0.9 | 10 13/16 |
| 1.0 | 12 |

| Inches | Decimal Feet |
|:---:|:---:|
| 1 | 0.083 |
| 2 | 0.17 |
| 3 | 0.25 |
| 4 | 0.33 |
| 5 | 0.42 |
| 6 | 0.50 |
| 7 | 0.58 |
| 8 | 0.67 |
| 9 | 0.75 |
| 10 | 0.83 |
| 11 | 0.92 |
| 12 | 1.00 |

## Fraction and Decimal Conversions

| Fraction | Decimal |
|:---|:---:|
| 1/16 | 0.0625 |
| 2/16, 1/8 | 0.125 |
| 3/16 | 0.1875 |
| 4/16, 2/8, 1/4 | 0.25 |
| 5/16 | 0.3125 |
| 6/16, 3/8 | 0.375 |
| 7/16 | 0.4375 |
| 8/16, 4/8, 2/4, 1/2 | 0.5 |
| 9/16 | 0.5625 |
| 10/16, 5/8 | 0.625 |
| 11/16 | 0.6875 |
| 12/16, 6/8, 3/4 | 0.75 |
| 13/16 | 0.8125 |
| 14/16, 7/8 | 0.875 |
| 15/16 | 0.9375 |
| 16/16 | 1 |

## Thread Protector Cap Colors

| Sizes | Examples | Rigid Color | IMC Color |
|:---|:---|:---:|:---:|
| Inch sizes | 1″, 2″, 3″, 4″, 5″, 6″ | Blue | Orange |
| ½″ sizes | ½″, 1½″, 2½″, 3½″ | Black | Yellow |
| ¼″ sizes | ¾″, 1¼″ | Red | Green |

# Conduit Bending and Fabrication Products

## A.J. Bender, LLC

3439 Northboro Court
Murfreesboro, TN 37129
(615) 494-9388
www.ajbender.com      *Propane PVC benders*

## Carlon

25701 Science Park Drive
Cleveland, OH 44122
(216) 464-3400
www.carlon.com      *PVC conduit and fittings*

## Contractor's Choice, Inc.

2070 Schappelle Lane
Cincinnati, OH 45240
(800) 670-8665
www.contractorschoiceinc.com      *Power tools*

## Cooper B-line

509 West Monroe Street
Highland, IL 62249
(800) 356-1438
www.cooperbline.com      *Conduit support systems*

## Erico, Inc.

34600 Solon Road
Solon, OH 44139
(800) 853-0878
www.erico.com      *Conduit supports*

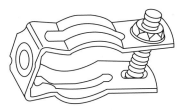

*Conduit bending tools*

## Gardner-Bender

6101 N Baker Road
Milwaukee, WI 53209
(800) 822-9220
www.gardnerbender.com

*Conduit benders, hole saws,
punches, and tools*

## Greenlee, A Textron Company

4455 Boeing Drive
Rockford Illinois 61109
(800) 435-0786
www.greenlee.com

*PVC propane bending torches*

## Hot Bend Corp

PO Box 202183
Arlington, TX 76006
(877) 468-2363
www.hotbend.com

*Conduit fittings*

## Hubble-Raco

3902 West Sample Street
South Bend, IN, 46619
(800) 722-6437
www.hubbell-raco.com

*Hand benders and tools*

## Ideal Industries, Inc.

3902 West Sample Street
South Bend, IN, 46619
(800) 435-0705
www.idealindustries.com

*Levels*

## Johnson Level and Tool Mfg. Co., Inc.

6333 W Donges Bay Road
Mequon, WI 53092
(262) 242-1161
www.johnsonlevel.com

## Jim Kerry

P.O. Box 38125
St. Louis, Mo. 63138
314-288-4967

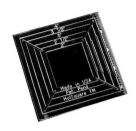

*HOLSquare™ conduit entry template*

## Klein Tools

7200 McCormick Blvd
Skokie, IL 60659
(800) 553-4676
www.kleintools.com

*Hand benders and tools*

## Lidseen of North Carolina

PO Box 207
Hayesville, NC 28904
(800) 742-3538
www.chicagobender.com

*Mechanical benders*

## Maxis

1225 W Houston
Suite 103
Gilbert AZ 85233
(888) 266-2947
www.maxis-tools.com

*Layout tools*

## MEGAPRO

Unit 102, 360 Edworthy Way
New Westminster, B.C., V3L 5G5, Canada
(866) 522-3652
www.megapro.net

*Screwdrivers*

## Milwaukee Electric Tool Corp

13135 W Lisbon Road
Brookfield, WI 53005
(800) 729-3878
www.milwaukeetool.com

*Power tools*

### Ridge Tool Co.

400 Clark St
Elyria, OH 44035
(888) 743-4333
www.ridgid.com

*Conduit fabrication tools*

### Ron Aubrey, Inc.

520 N. 3rd St.
Nicholsville, KY 40356
(859) 885-7317
www.no-dog.com

*No-Dog offset level*

### Snap-On Tools

2801 80th Street
Kenosha, WI 53143
(262) 656-5200
www.snapon.com

*Hand tools*

### Steel Tube Institute of North America

2000 Ponce de Leon Suite 600
Coral Gables, Florida 33134
(305) 421-6326
www.steeltubeinstitute.org

*Trade group representing manufacturers of steel tube and conduit*

### Stout Tool Corporation

29233 Haas Rd # A
Wixom, MI 48393
(877) 337-8688
www.stouttool.com

*Battery-powered band saws*

### Thomas & Betts Corp

8155 T&B Blvd, MS 3A-16
Memphis, TN 38125
(800) 816-7809
www.tnb.com

*Conduit fittings and copper shield*

Further information may be found by clicking on the Reference Material button on the CD-ROM home screen.

# Glossary

## A

**angle:** A measure of the rotation between two lines that are joined together at one point.

## B

**bender foot pedal:** The part of the bender where foot pressure is applied in order to bend the conduit.

**bender handle:** A tube or lever used to hold the bender while in use.

**bender hook:** The part of the bender shoe that holds the conduit in place during the bending process.

**bender shoe:** The curved part of a bender that forms the conduit during fabrication.

**box offset:** A small offset bend used to move conduit away from a wall and into an electrical box.

## C

**chain vise:** A tool that consists of a chain that wraps around the conduit and a tightening screw that secures the chain around the conduit.

**closed offset bend:** An offset where the second bend is made with an angle that is too small and the end of the conduit falls relative to the top of the obstacle.

**concentric bend:** Two or more segmented bends nested together.

**conduit reamer:** A tool used to remove burrs and sharp edges from a piece of conduit after it has been cut to length.

**conduit threader:** A tool that is used to cut threads in conduit.

**corner offset:** A bend consisting of two offsets turned at a 90° angle from each other.

## D

**datum:** A reference point to which other elevations, angles, or measurements are related.

**die:** A cutting tool used to form external screw threads in conduit.

**die head:** The part of a conduit threader that holds the dies securely in position and applies pressure to cut external threads.

**dogleg:** A multiple bend in conduit where one of the bends is not in the same plane as the other bend.

**drophead threader:** A ratchet threader in which the entire die head can be replaced with another die head for different conduit sizes.

## E

**electric bender:** A type of bender that uses an electric motor to rotate a series of shoes used to fabricate conduit bends.

**electrical metallic tubing (EMT):** A lightweight tubular steel raceway without threads on the ends.

## F

**four-bend saddle:** A saddle made by placing four bends in a conduit to allow it to bend around an obstacle and then return to its original level.

## G

**gain:** The difference between the sum of the straight distances and the actual length of conduit.

## H

**hickey:** A hand bender with a no-radius shoe.

**hot box:** A PVC heating tool containing an electric heating element within an enclosure that holds the heat.

## I

**intermediate metal conduit (IMC):** A raceway of circular cross-section with an intermediate wall thickness designed for protection and routing of conductors.

## K

**kick:** Any bend of less than 90° that is used to change direction in a conduit run.

**mechanical bender:** A type of bender that employs a lever arm and ratcheting mechanism to provide a mechanical advantage when fabricating conduit bends.

**National Pipe Taper (NPT) thread:** A standard thread used for connecting conduit in which the adjoining sides of the threads are at a 60° angle to each other.

**nonratcheting threader:** A conduit threader in which the handles are rotated completely around the conduit to turn the die head.

**offset bend:** A double conduit bend with two equal angles bent in opposite directions in the same plane in a conduit run.

**oiler:** A device used to apply cutting oil to the dies during the thread cutting process.

**open offset bend:** An offset where the second bend is made with an angle that is too large and the end of the conduit rises relative to the top of the obstacle.

**push-through method:** Any procedure for bending conduit in which the conduit is not turned around end for end during the bend.

**quick-opening die head:** A die head with a release lever at the top that can be raised manually to retract the dies and release the conduit after the thread has been cut.

**ratchet threader:** A conduit threader with a ratchet mechanism built into the handle and that requires only a small amount of clearance for proper operation.

**reamer:** A tool used to remove burrs and sharp edges from a piece of conduit after it has been cut to length.

**right angle:** An angle that measures exactly 90°.

**right triangle:** A triangle where one of the angles is a right angle.

**rigid metal conduit (RMC):** A threadable conduit with fairly thick walls.

**rigid nonmetallic conduit (RNC):** A conduit made of materials other than metal.

**rule:** A rigid measuring tool.

**S**

**saddle:** A section of conduit consisting of three or four bends that is shaped to bend around an obstacle and then return to its original level.

**segmented bend:** A bend that consists of a series of small bends made at predetermined locations on a piece of conduit to create one large bend. Each small bend is called a shot.

**shrink constant:** The reduction in distance that a conduit can run per inch of offset elevation.

**shrink:** The amount by which the total run that conduit can cover is reduced because of the extra length required to bend around an obstacle.

**springback:** The property of conduit that causes it to unbend slightly after a bend is completed.

**stub-up bend:** A 90° bend in conduit, made perpendicular to the original length of the conduit, with the conduit extending a specified length from the back of the bend.

**T**

**take-up:** The value that is used to determine where to place the bending marks.

**tape measure:** A measuring device with a metallic tape wound in a coil that can be extended to take measurements.

**three-bend saddle:** A saddle consisting of a center bend and two side bends, with the center bend having twice the angle of the side bends.

**three-way threader:** A nonratcheting threader that holds three die sizes simultaneously.

**trigonometry:** The branch of mathematics used to determine the sides and angles of triangles.

**wow:** *See dogleg.*

**Y**

**yoke vise:** A tool that consists of a tightening screw and a yoke holding a set of jaws that are used to securely hold the conduit.

# Index

## A

adjacent side, 31
admixture, 170
advanced bending techniques. *See*
 concentric bends; segmented
 bends
aluminum conduit, 116
 plastic thread protector caps, *116*
angles, 30, *31*
 bend angle, *33, 141*
  completed bend angle, 141
 calculating shrink with
  nonstandard angles, 46
 reference angle, 31
 right angle, 30, *31*
 theta (θ), 31
anti-dog (no-dog) level, 78, *79*
applications
 arched ceiling, 152
 compound 90° bend around a
  sprinkler, 72, *73*
 conduit rack with concentric
  bends, *148*, 148–152
 control boxes mounted on
  cylindrical heating tank, *146*
  measuring straight distance, *147*
 enclosures mounted on pressure
  vessel, 144, *145*
 offset around duct, 37
 offset around duct with shrink, 45
 parallel offsets on a conduit rack,
  48–49
 parallel offsets with conduit of
  different sizes, 54–55
 three-bend back-to-back bends, 25
 three-bend saddle around a drain,
  73–74, *74*

## B

backfilling, 171–172
 compaction, 171–172, *172, 173*
 materials, 171

back-to-back 90° bends, 21, *23,
 24, 25*
 layout and fabrication, 26
 three-bend back-to-back bends,
  25–27
 three-bend gain, 26
benchmarks, 157
 datum, 157
bend angles, 60, 69, 71
 90° bends, 17–18
 choosing, *33*, 33–34
 saddle bends, 60
bend corrections, 20, 38
bend errors, 38, *39*
 closed offset bend, 38
 dogleg, 38, *40*
 open offset bend, 38
benders. *See also* hand benders;
 electric benders; mechanical
 benders; hydraulic benders
 components, 13
bender shoe, *13*
 markings, *14*

bends
 back-to-back 90° bends, 21, *23*
  three-bend back-to-back bends,
  25
 basic, 17–21
 calculating the distance between
  bends, 34, *87*, 102–103
  distance multiplier and shrink
  constant, *56*
 choosing the bend angle, 33–34
 closed offset bend, 38
 concentric bends, 147
  arched ceiling, 152
  concentric bend dimensions, *149*
  conduit layout, *151*
  conduit lengths, *150*
  conduit rack with concentric
   bends, *148*, 148–152
 dogleg, 21
 pre-positioning, 23

 segmented, 140–147
  control boxes mounted on
   cylindrical heating tank,
   *146, 147*
  enclosures mounted on pressure
   vessel, 144, *145*
  layout, *143*, 143–144, *151*
  number of, 142
  tips for making, 153–154
  variables, 140–142
bend pre-positioning, 23
 gain, 23, *25, 26,* 27
 measuring, 24
B method, 22
box offsets, *40*
brass, 116

## C

chain vise, 128
charting a bender, 82–86, *83–84*
 gain, 83, *84*
 radius adjustment, *84*, 84–85
 setback, 83
 take-up, 82, *83*
 travel, 85
 travel per degree, 86
Chicago benders. *See* mechanical
 benders
closed offset bend, 38
close nipple chuck, *132*
compaction, 171–172, *172, 173*
compound 90° bends, *70*, 70–73
 application
  compound 90° bend around a
   sprinkler, 72, *73*
 calculated layout method, 71
  distance between bends, 71
 layout, *73*
 measured layout method, 70–71
  distance between bends, 70–71
 measurement, *71*

concentric bends, 147–152
  applications
    arched ceiling, 152
    conduit rack with concentric
      bends, *148,* 148–152
  dimensions, *149*
  conduit lengths, *150*
  conduit layout, *151*
  conduit layout lengths table, *151*
concrete, 170–171
  components, 170
    admixture, 170
    Portland cement, 170
  testing, 171
    slump test, *171*
  types, 170
conduit (history)
  early conduit designs, 3–4
    early wiring methods, 2
    iron gas pipe, 3
    spiral-wound paper tubes, 3
    steel conduit, 4
    zinc tubes, 3
  Edison, Thomas, 2, *3*
  Pearl Street Station, 2, *3*
conduit and the NEC, 4–8
  electrical metallic tubing, 5, 6
  intermediate metal conduit, 8
  rigid metal conduit, 4–5, *5*
    thread protector caps, *5*
  rigid nonmetallic conduit, 7–8
  PVC, *7*
conduit length
  calculations, 26
conduit reamer, 12, 15
conduit support, *132*
conduit threaders, 126–132
  dies and die heads, 126, *127*
    large conduit die heads, *135*
  oilers, 126, *128*
  reamers, 127, *128*
  handheld threaders, 128–131, *130*
    hand-driven threaders, 130
    nonracheting threaders, 128,
      *129*
    power-driven threaders, *130,*
      130–131
    racheting threaders, 128, *129*
    vises, 128
  large power threaders
    power vises, *129,* 131, 132, 136
  threading tools, *126*
converting between fractions and
    decimals, *16,* 16–17
corner offsets, *69,* 69–70
  making the offset bends, 70
  marking the bend locations, 69–70
cosecant (csc), 31–32
cosine (cos), 31–32
cotangent (cot), 31–32

**D**

datum, 157
deduction. *See* take-up
die, 126
die head, 126
  large conduit die heads, *135*
distance multiplier, *56*
distance to obstruction, 59
dogleg, 21, 38, *40,* 71, *79*
drophead threader, 128

**E**

Edison, Thomas, 2, *3*
electrical metallic tubing, 5–7, *6,* 110
electric benders, *99,* 99–101
  components, 99, *100*
  kicks, 100–101
    measured rise method, 101
  operation, 99–100, *101*
electronic distance measurement
    (EDM), 160
EMT. *See* electrical metallic tubing
excavation, 164–169, *165*
  safety, 164–167
  shielding, 169
    trench box, 169
  shoring, *168,* 168–169
  sloping and benching, *167,*
    167–168

**F**

four-bend saddle, 58, 66–68, *68*
  layout, *67*
  making the offset bends, 68
  marking the bend locations, 66
    shrink calculation, 66
  measurements, 59
  shrink, 62
functions, trigonometry, *31,* 31–32

**G**

gain, 23, *25,* 26, 27
  back-to-back gain, 24–25
  measuring, 24
global positioning system (GPS), 160
glycol heaters, 122
grout, 170

**H**

hand benders, *12,* 12–14
  bender foot pedal, 13
  bender handle, 13
  bender hook, 13
  bender shoe, 13
  bender shoe markings, 14
  hand bending gain, *25*
  plumb 30 and plumb 45 benders, 12
hand bending. *See* corner offsets;
    kicks; offset bends; saddle
    bends
hickeys, 12, 14, *15,* 17
hot box, 122, *123*
hubs, *157. See* benchmarks
hydraulic benders, 105–114
  bender components, *108*
    multiple-shot benders, 107, *108*
    one-shot benders, 107, *108*
  set up for layout and fabrication,
    109–113
    layout, 109
    offsets and kicks, 111–114
    ram travel, 109–110, 112
hydraulics, 106
  Pascal's law, 106
  system components, 106, *107*
  system safety, 106
hypotenuse, 31, 42–43

**I**

IMC. *See* intermediate metal
    conduit
incandescent lamp, *2*
intermediate metal conduit, 2, 8, 110
  thread protector cap, 8
iron gas pipe, 3–4

## K

kicks, *52,* 52–54
  electric benders, 100–101
  hydraulic benders, 111–114
    bending kicks, *113*
  measured rise method for kicks,
    52, *53,* 90–91, *91*
  mechanical benders, 88–91
  multiplier method for kicks,
    52–54, *54,* 88–90, *90*

## L

layout, 81–87
  charting a bender, 82–86, *83–84*
leveling rods, 163
levels
  automatic (self-leveling), 158
  builder's, 158
  engineer's, 158
  instrument height, *159*
  level horizontal circle, *159*
  mirrored, 161
  spirit, 161

## M

marking the bend locations, 34
measured rise method for kicks, 52,
    *53,* 90–91, 113
  electric benders, *91,* 101
measured rise method for offset
    bends, *50,* 50–54, 111–112
  making the bends, 50–51
  second bend, *51*
measuring ram travel, *110*
measuring shortcuts, 20, *21*
mechanical benders, *78,* 78–98
  bending aids, 78–80
    no-dog level, *78, 79,* 80
  components, 78, *79*
    HVAC components, *79*
  fabrication, 88–97
    kicks, 88–91
    90° bends, 88, *89*
    offsets, 91–93, *93*
    saddle bends, 94–97
  layout, 81–87
    charting a bender, 82–86, *83–84*
    pre-positioning bends, 86–87
    push-through method, 82, 86–87
  operation, 80, *81*

metallic conduit, 116
  aluminum conduit, 116
    plastic thread protector caps, *116*
  brass, 116
  silicon-bronze, 116
  stainless steel, 116
mortar, 170
multiple-shot benders, 107, *108*
multiplier method for kicks, 52–54,
    *54, 90,* 113
multiplier method for offset bends,
    32–37
  application
    offset around duct, 37–38
  calculating the distance between
    bends, 32
  choosing the bend angle, 33–34
  distance between bends, *34*
  making the first bend, *36,* 37
  making the second bend, 37
  marking bend locations, *35*
  measuring offset rise, 32

## N

National Pipe Taper (NPT) thread,
    133, *134*
90° bends, 17–18
  back-to-back 90° bends, 21, *24,*
    *25*
    measurements, *23*
    three-bend gain, *26*
  making bends at right angles, 22,
    *24*
  with mechanical benders, 88
  one-shot benders, 110, *111*
  reverse method, 22
  stub-up, 18, *19, 20*
no-dog level, *78, 79,* 80
nonratcheting threader, 128, *129*

## O

obstruction rise, 59, *72*
offset bends, *30,* 30–56, *38*
  bend corrections, 38
  bend errors, 38, *39*
  box offsets, *40*
  measured rise method, *50,* 50–54
    making the bends, 50–51
    second bend, *51*
  multiplier method, 32–37
    calculating the distance between
      bends, 34

choosing the bend angle, 33–34
    marking the bend locations, 34,
      *36*
    measuring the offset rise, 32
  parallel offset bends, *47,* 47–49,
    *48, 49,* 55
  pre-positioning, 40–47
    calculating shrink, 41, 43
    finding precise offset locations
      using shrink, 43
    offset around duct with shrink, 45
  rolling offsets, 42–43
offsets, *38*
  hydraulic benders, 111–114, *112*
  mechanical benders, 91–93, *93*
    fabricating offset bends, 92–93
    offset layout, 92
  offset measurement, *33*
    distance multiplier and shrink
      constant, *56*
oiler, 126, *128*
one-shot benders, 107, *108*
  90° bends, 110, *111*
open offset bend, 38
opposite angle, 31

## P

parallel offset bends, *47,* 47–49,
    *49,* 55
  applications, 54
    parallel offsets on a conduit
      rack, 48–49
    calculating parallel offset
      adjustments, *48,* 55
Pascal's law, 106
pawl release, 80
Pearl Street Station, 2, *3*
plumb benders, 12
plumb bobs, 161, *163*
polyvinyl chloride. *See* PVC
Portland cement, 170
power ponies, 130, *136*
printreading, 157–158
  abbreviations, *158*
push-through method, 65, 81–87
PVC-coated conduit, 117–120
  coating, 117
    bare steel, 117
    coating repair, 117
    galvanized steel, 117
    outside diameter, *117*
    zinc-coated steel, 117
  conduit sizes, 117

couplings and fittings, 119–120
  modified tools, *120*
  sleeves, 119
tools, 117–119
  bender shoes, *119*
  bending, 119
  clamping, 118
  cutting and threading, 118, *119*
PVC conduit, 7, 120–122
  bending, 121–122
  heaters, 122–124
    glycol, 122
    hot box, 122, *123*
    pipe plugs, *122*
    torches, 122
  joining, 120–121
    bending with a template, *121*
    solvent, *120*
    cutting, *121*
  sizes, 120
Pythagorean theorem, 42–43

**Q**

quick-opening die head, 126

**R**

raceways and conduit, 2–10
  early conduit designs, 3–4
    early wiring methods, 2
    iron gas pipe, 3
    spiral-wound paper tubes, 3
    steel conduit, 4
    zinc tubes, 3
  electrical metallic tubing, 5–7, *6*
  intermediate metal conduit, 2, 8
    thread protector caps, *5*, 8
  rigid metal conduit, *5*
  rigid nonmetallic conduit, 2, 7–8
    permitted uses, 7
    uses not permitted, 7–8
ratchet threader, 128, *129*
ram travel, 109–110, 112
  predetermined ram travel, *110*
reamer, 127, *128*
right triangle, 30–31, *31*
  adjacent side, 31
  hypotenuse, 31, 42–43
  opposite side, 31
  reference angle, 31
rigid. *See* rigid metal conduit
rigid metal conduit, 2, 4–5, *5*

thread protector caps, 4, *5*
rigid nonmetallic conduit, 2, 7–8
  permitted uses, 7
  uses not permitted, 7–8
RMC. *See* rigid metal conduit
rolling offsets, 42–43
  Pythagorean theorem, 42–43
rules
  folding rules, 15

**S**

saddle bends, 58–76
  application
    three-bend saddle around a
      drain, 73–74
  four-bend saddle, 58, 66–68, *98*
    layout, *67*, *96*
    making the offset bends, 68
    marking the bend locations, 66
    measurements, 59
    shrink, *62*
  mechanical benders, 94–97
    four-bend saddles, 95–97, *96, 98*
    three-bend saddles, 94, *95*
  shrink calculation, 60–61
    shrink for a cylindrical
      obstruction, 61
    shrink for a rectangular
      obstruction, 61
  three-bend saddle, 58, 61–64, *95*
    layout, *63, 74, 94*
    making the bends, 63–64, *64*
    measurements, 59
    push-through method, 65
    value of the distance multiplier,
      *63*
  variables, 59–60
    bend angles, 60
    distance to obstruction, *59*
    obstruction rise, 59
safety, excavation, 164–167
  hydraulic system safety, 106
secant (sec), 31–32
segmented bends, 140–147
  application
    control boxes mounted on
      cylindrical heating tank, *146*
    enclosures mounted on pressure
      vessel, 144, *145*
    layout, *143*, 143–144, *151*
    even number of shots, 143–144,
      *144*
    layout lengths table, *151*

odd number of shots, 143, *144*
  tips for making, 153–154
  variables, 140–142
    bend radius, *140*, 140–141
    completed bend angle, 141
    developed length, *141*, 141–142
    number of segmented bends, 142
shielding, 169
  trench box, 169, *172*
shoring, *168*, 168–169
shrink, *41*, 87, 102
  calculating shrink, *41*, 43–46,
    60–61
    for cylindrical obstructions, 61
    for nonstandard angles, 46
    for rectangular obstructions, 61
    for standard angles, *41*
    four-bend saddles, *62, 66, 67*
    three-bend saddles, *61*
  shrink constant, *41, 56*, 60–61, *66*
    distance multiplier and shrink
      constant, *56*
shrink constant, 41, *56*, 60–61, *66*
silicon-bronze, 116
sine (sin), 31–32
sleeves, 119
sloping and benching, *167, 168*–169
springback, 13, 110
stainless steel, 116
stub-up, 18, *19*, 20, 21
surveying instruments, 158
  electronic distance measurement
    (EDM), 160
  global positioning system
    (GPS), 160
  leveling rods, 163
  levels
    automatic (self-leveling)
      levels, 158
    builder's levels, 158
    engineer's levels, 158
    instrument height, *159*
    level horizontal circle, *159*
    mirrored levels, 161
    spirit levels, 161
  plumb bobs, 161, *163*
  transits and theodolites, 160
  tripods and leveling, 161
    tripod mounts, *161*

**T**

take-up, 17–18, *18*, 20, 22
tangent (tan), 31–32
tape measures, 15

Tesla, Nikola, 2
theta (θ), 31
threaded conduit, 133–136. *See also* conduit threaders
  procedure, 134–136, *135*
    threading large conduit, 136
  threading tools, *126*
  thread length and taper, 133
    National Pipe Taper (NPT) thread, 133, *134*
threaders. *See* conduit threaders
thread protector caps, 4, *5*, 8
three-bend saddle, 58, 61–64
  application
    drainpipe layout, *74*
    three-bend saddle around a drain, 73–74, *74*
  making the bends, 63–64
    center bend, 63, *64*
    second and third bends, 64
  marking the bend locations, 61–63
    marking the center bend, 62, 75
    marking the outside bends, 62–63
    value of the distance multiplier, *63*
  measurements, 59
  three-bend saddle layout, *63*
  three-bend saddle shrink, *61*
three-way threader, 128
tools, 12–15
  calculators, 12, *16*, 16–17
  conduit reamer, 12, *15*
  die, *126*, *127*
  hand benders, 12–14
  hickeys, 12, 14, *15*
  levels, 12, 21
    torpedo level, 13
  measuring tape, 20
  no-dog level, 78
  protractors, 12, 13
  PVC-coated conduit, 117–119
    bender shoes, *119*
    bending, 119
    clamping, 118
    couplings and fittings
    cutting and threading, 118, *119*
  reamer, 127, *128*
  rules, 12, 15
  tape measures, 12
  threading tools, 126
  torpedo level, *13*
  vises, 128, *129*, 131, 136
torch, 122
torpedo level, 13, 18

transits and theodolites, 160
travel, 85
trench box, 169, *172*
trigonometry, 30
  functions, *31*, 31–32
tripods. *See* surveying instruments

**U**

underground conduit installation, 155–174
  backfilling, 171–172
    compaction, 171, *172*, *173*
    materials, 171
  excavation, 164–169, *165*
    excavation safety, 164–167
      cave-ins, 165, *166*
      confined spaces, *164*, *165*
      ingress and egress, 166, *167*
      water accumulation, 166–167
    shielding, 169
      trench box, 169
    shoring, *168*, 168–169
    sloping and benching, *167*, 167–168, *168*
  job planning, 156–158
    benchmarks, 157
      datum, 157
      hubs, *157*
    printreading, 157–158
      abbreviations, *158*
    specifications and site plans, *156*, 156–157
  procedures, 169–171
    concrete, 170–171
      admixture, 170
      components, 170
      Portland cement, 170
      slump test, *171*
      testing, 171
      types, 170
  surveying instruments, 158
    electronic distance measurement (EDM), 160
    global positioning system (GPS), 160
    leveling rods, 163
    levels
      automatic (self-leveling) levels, 158
      builder's levels, 158
      engineer's levels, 158
      instrument height, *159*
      level horizontal circle, *159*

    mirrored levels, 161
    spirit levels, 161
    plumb bobs, 161, *163*
    transits and theodolites, 160
    vertical arc, *160*
    tripods and leveling, 161
      tripod mounts, *161*

**V**

vises, *129*
  chain vise, 128
  yoke vise, 128
  power vises, 131
    accessories, 132
    components, *132*

**W**

wow. *See* dogleg

**Y**

yoke vise. *See* vises

## USING THE *CONDUIT BENDING AND FABRICATION* CD-ROM

*Before removing the CD-ROM from the protective sleeve, please note that the book cannot be returned for refund or credit if the CD-ROM sleeve seal is broken.*

### System Requirements

The *Conduit Bending and Fabrication* CD-ROM is designed to work best on a computer meeting the following hardware/software requirements:

- Intel® Pentium® (or equivalent) processor
- Microsoft® Windows® 95, 98, 98 SE, Me, NT®, 2000, or XP operating system
- 64 MB of free available system RAM (128 MB recommended)
- 90 MB of available disk space
- 800 × 600 16-bit (thousands of colors) color display or better
- Sound output capability and speakers
- CD-ROM drive
- Internet Explorer™ 3.0 or Netscape® 3.0 or later browser software

### Opening Files

Insert the CD-ROM into the computer CD-ROM drive. Within a few seconds, the home screen will be displayed allowing access to all features of the CD-ROM. Information about the usage of the CD-ROM can be accessed by clicking on USING THIS CD-ROM. The Quick Quizzes™, Illustrated Glossary, Bending Calculator, Procedural Videos, and Reference Material can be accessed by clicking on the appropriate button on the home screen. Clicking on the American Tech web site button (www.go2atp.com) accesses information on related educational products. Unauthorized reproduction of the material on this CD-ROM is strictly prohibited.